Innovations in Food Processing Industry

NIPA® GENX ELECTRONIC RESOURCES & SOLUTIONS P. LTD.
New Delhi-110 034

About the Editors

Dr. Dev Raj is Professor and Head (Post Harvest Technology), in the Department of Post Harvest Technology, ASPEE College of Horticulture, Navsari Agricultural University, Navsari, Gujarat India. He also worked as Scientist (Food Technology), in the Department of Food Science and Technology, Dr YS Parmar University of Horticulture and Forestry Nauni, Solan (HP) 173230, India. He is having more than 20 year's experience in research, teaching and extension experience in the field of Post Harvest Technology and Food Science & Technology. He is graduated in Horticulture from Dr YS Parmar University, Solan; MSc in Post Harvest Technology of Horticultural Crops from University of Agricultural Sciences, GKVK, Bangalore and PhD in Post Harvest Technology of Horticultural Crops from Dr YS Parmar University, Solan. He has been awarded with ICAR Junior Research Fellowship during MSc and merit Scholarship during PhD. He also served at Directorate of Mushroom Research (Indian Council of Agricultural Research), Chambaghat as Senior Research Fellow for about 2 years. He has a very good experience in the field of processing, preservation and value addition of fruits and vegetables. He has to his credited more than 70 research papers published in journals of national and international repute along with more than 70 popular articles/ technical articles, 17 book chapters, 11 books related to Food Processing. He was Co-PI and project leader for different projects under All India Coordinated Research Project on Post Harvest Technology (ICAR). He also handled externally funded project of NOVOD Board, GOI as Co-PI. He has more than 30 research recommendations for the farming communities. He handled more than 30 research experiments as PI. He has guided 9 MSc and 8 PhD Students as major guide. He is recipient of various awards including Fellow award of ISHRD, Kejriwal award (2008) and Dr J.S. Pruthi Memorial Award (2010) for their research contribution in the interest of food industries developmentby All India Food Processor's Association, New Delhi. He is member for many professional societies. He is referee of prestigious Journal of Food Science and Technology. He is also appointed as paper setter cum evaluator by many agricultural universities.

Dr. Z. P. Patel is **Vice-Chancellor** of Navsari Agricultural University, Navsari (Gujarat). He born on 1st May, 1957, native of Navsari, Gujarat has did his B.Sc. (Zoology) from South Gujarat University, Surat; M.Sc (Agri.) in Agricultural Entomology from erstwhile Gujarat Agricultural University, Navsari and Ph.D in Agricultural Entomology from GAU, Anand (Gujarat) in 1994. He reached to the highest position of Vice-Chancellor, after serving various positions viz. JRF, SRF, Senior Research Assistant, Assistant Professor, Associate Professor, Professor. He was first Principal of College of Agriculture, Waghai and worked as Dean, Faculty of Agriculture at NAU Navsari. He also established accredited NABL Food Quality Testing Laboratory at Navsari during his tenure as Professor. For all these efforts, he is undoubtedly titled as ***'Man of Infrastructure Development'*** in the University. He possessed about 40 years of experience in Research, Education, Extension & Administration and 27 years' hands-on experience in Tribal belt. As a Vice Chancellor, now he has taken the Name and Fame of the Navsari Agricultural University to new heights by improving the standards of research, teaching, extension, infrastructure development and administrative improvements in every aspect. He had Major contributions in the Pest Management; and has important research achievements in the field of Entomology and worked specially to control Fruit flies in Fruit crops and Vegetables. He developed and commercialized one kind of first fruit fly trap (**Nauroji Stone House Fruit Fly Trap**) in the country which is very effective, economic, eco-friendly and sustainable for the management of fruit flies enhancing the export potential and is patented technology. His work on fruit fly gave him recognition as **"Man of Fruit Fly"** in academic and farming fraternity. He also developed **Crop Pest Map** in Sapota which guided farmers to implement Integrated Pest Management (IPM) strategies more efficiently in time to protect the crop and reap higher income. His contribution for control of **Sapota Seed/Fruit Borer;** an insect named *Trymalitis margarias* found first time in the world is of paramount importance and saved the crop in Gujarat as well as in country from its destruction. He is also known for his contribution for **Bio-Control of Sugarcane Pyrilla** which was one of the most devastating pest of sugarcane in South Gujarat and has been controlled by a very effective natural enemy, *Epiricania melanoleuca* which is proved as most successful example of Bio-control in Gujarat. He has guided total **17** students (12 M.Sc & 5 Ph.D) in Agricultural Entomology as Major advisor. He is also a good extension educationist and wholeheartedly participated in Krishi Mahotsav and successfully convinced farmers for the adoption of new technologies. He also delivered numerous TV talks and radio talks. He was successful in creating large scale awareness among farmers regarding use of fruit fly trap. Dr. Patel has in his credits about more than 5 patents, 47 research papers published in referred international and national journals, 03 Books published, 04 papers presented in International seminar, 12 paper presented in National seminar, 10 Research technologies developed for farmers, 38 popular articles and booklets, 02 lead lectures delivered in National conference, 04 projects completed as Principle Investigator and 06 projects initiated as Principal. He has acted as Chairman/Member/Convenor in many research, extension, academic forums of NAU, Navsari before joining as Vice-Chancellor NAU, Navsari. During the tenure of his services he received Awards, recognitions and appreciations and is a proud recipient of the most coveted Research Excellence Awards from two consecutive Hon. Chief Ministers of Gujarat State (i) First by Shri Keshubhai Patel during 2000 (Sardar Patel Agricultural Research Award, GoG) for the Crop-Pest map in Sapota (ii) Shri Narendrabhai Modi in 2012 (Federation of Gujarat Industries Award) for the Fruit fly trap. Other awards include **Significant Achievement Award** by Rotary International, Gandevi in 2001 and Publication of success story "Integrated Pest Management for Higher Fruit Production" of RKVY in "Coffee Table Book" – a publication of GoI in 2012. He also has foreign exposure as attended 25 days international training at Israel and delivered lecture in UC Davis University of California, USA. He is always a man of different thoughts and his *Jarahatke approach* made him a versatile person in all his work. He is a good team leader, punctual, visionary, and a quick decision-maker. A true son of the soil, Dr. Z. P. Patel has always thought for the welfare of farmers and students; and his untiring efforts, due diligence, and perseverance have set an example for the scientific and teaching fraternity.

Innovations in Food Processing Industry

Edited by

Dev Raj
Professor & Head (PHT)
Department of Post Harvest Technology
ASPEE College of Horticulture
Navsari Agricultural University
Navsari-396450, Gujarat (India)

Z.P. Patel
Hon'ble Vice Chancellor
Navsari Agricultural University
Navsari-396450, Gujarat (India)

NIPA® GENX ELECTRONIC RESOURCES & SOLUTIONS P. LTD.
New Delhi-110 034

NIPA® GENX ELECTRONIC
RESOURCES & SOLUTIONS P. LTD.

101,103, Vikas Surya Plaza, CU Block
L.S.C.Market, Pitam Pura, New Delhi-110 034
Ph : +91 11 27341616, 27341717, 27341718
E-mail: newindiapublishingagency@gmail.com
www: www.nipabooks.com

For customer assistance, please contact
Phone: + 91-11-27 34 17 17
Fax: + 91-11-27 34 16 16

Print ISBN: 978-93-58875-66-9
ebook ISBN: 978-93-58879-34-6

Composed and Designed by NIPA®.

Preface

Food processing is the application of techniques in a systematic manner for preventing losses through preservation via processing, packaging, storage and distribution to ensure greater availability of a wide variety of foods.

It is a set of methods and techniques used to transform raw material into processed food for consumption by humans or animals. Keeping the importance of the food in our life, it is very important that all people either engaged in food processing or not, should know about the importance of food processing in decreasing the post harvest losses of perishable horticultural commodities. But to understand the processing and its importance in decreasing post harvest losses of food crops, some sort of documentation is needed. A manuscript entitled "**Innovations in Food Processing Industry**" contains all important information related to innovative techniques being used in food processing industries. The book is comprised of 21 different chapters. Each chapter in the book has been well illustrated for better understanding with important MCQ at the end of chapter. We acknowledge from the core of our heart for the valuable guidance and support received from authors across the country and numerous literature who have contributed tirelessly in the field of food processing of horticultural produce. It is hoped that the book will be welcomed and its benefits will be availed by an increased number of students, researchers, teachers and all those who have interest in the subject. Although, every care has been taken by the authors while writing this text book even then spelling mistakes, overlapping; repetition of brief portion here and there could not be helped. Needless to say, errors and omissions are solely our. Constructive criticisms and suggestion by the readers are welcomed for improving the quality of manuscript. Reader's indulgence in this regard is highly solicited.

Editors

Contents

1

Innovative Food Processing Technologies

Dev Raj, A.K. Senapati, F.M. Sahu, Nilam V. Patel and Z.P. Patel

Navsari Agricultural University, Navsari, Gujarat-396 450, India

Introduction

The food processing industry ranks fifth in size in the country and employs 16 lakh workers which is 19% of the country's industrial labour. It accounts for 14% of the total industrial output with 18% of industrial GDP and 6.3% of countries GDP. So, food processing plays a vital role in India's prosperity. The Food Processing Industry sector in India is one of the largest sector in terms of production, consumption, export and growth prospects. Food processing industry broadly comprises primary processing category and the other categories embracing the secondary, tertiary and further stages of downstream processing. Primary food processing is a major industry with lakhs of rice-mills/ hullers, flour mills, pulse mills and oil-seed mills. There are several thousands of bakeries, traditional food units and fruit/vegetable/spice processing units in the unorganized sector. In the organized sector, there are over 820 flour mills, 418 fish processing units, 6600 fruit/vegetable processing units and 171 meat processing units. Indian processing industries has installed capacity of 21 lakh tones; however these industries actually processed 9.8 lakh tones of the produce. So the total capacity utilization of industries is <30%. There are 70% units in home/cottage/ small scale sector and 30% units in large scale sector with capacity of 250 tonnes per annum and 30 tonnes/hr, respectively. Large scale contributes 70% of the production. The principal crops and commodities which constitute the primary base for the food processing industry and considered important sub-sectors in food processing industries are illustrated in Fig. 1.

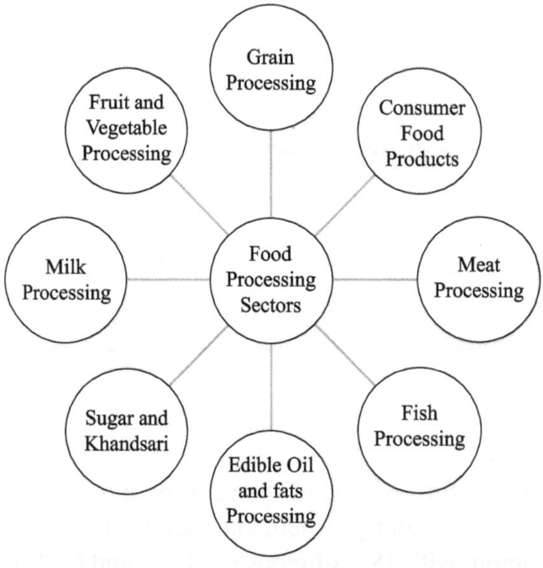

Fig. 1. Various food processing sectors available in the markets

In India majority of the produce is sent to the market for fresh trade and whatever is left as marketable surplus in the form of small and pittoo size, non-uniform and cull is utilized for processing for conversion into value added products. The processing capacity in our country from more than 36871 licensed Food Processing Units is less than 2.4% of total fruit and vegetable production against 60-70% in many advanced countries. Further, the actual production of processed products stood at only 1.33 million tones with a less than 35% capacity utilization of the installed processing units. There are 70% units in home/cottage/ small scale sector and 30% units in large scale sector with capacity of 250 tonnes per annum and 30 tonnes/hr, respectively. Large scale contributes 70% of the production. Despite such a huge production of horticultural commodities there is only <2.4% processing and India's share in the world trade is around 1%. However, processing in advanced countries is too high as compared to India. Status of processing in India is presented in Table 1 and 2 and Fig 2 & 3. India is the land of spices producing all varieties worth Rs. 3500 crores amounting to 25-30% of world production, which is processed for value addition and export. Tea, coffee, cashew, cocoa, sugarcane etc. are the major plantation crops in the country. India's principal crops accounted for 5-6% of the India's total export earnings. India is the world's largest producer of the tea. India is the largest producer (31% of world production), processor and exporter of the cashew kernels (48% of the world export). Consumer food comprises of group of products like chocolates, confectionery, cocoa products, soya based products, RTE foods, mineral waters, soft drinks etc. Among these, soft drinks enjoy the biggest share. Confectionery output is growing at the rate of 6-7% and chocolate production at the rate of 10-15%. Mango, grapes, apple, onion, potato, green vegetables like okra, bitter gourd and green chillies were the major items of export. Similarly, India exported 9.53 Lakh tones of processed foods comprising of mango pulp, juices, concentrates, dried and processed vegetables, pickle and chutney, alcoholic and non-alcoholic beverages worth Rs. 10583.41 crores (Table 3 and 4).

Table 1: Status of post harvest management and processing

Status	**Position**
Fruits and vegetables Processing	<2.4%
Export of Processed Products	9.53 Lakh tones
Value	Rs. 10583.41 Crores
Processing capacity	3.85 million tones
Actual Production	1.33 million tones
Capacity utilization	Less than 35%

Table 2: Status of processing units in different states of India

Sr. No.	**State**	**Total no. of Processing units**	**Sr. No.**	**State**	**Total no. of Processing units**
1	Andhra Pradesh	9359	16	Maharashtra	3113
2	Arunachal Pradesh	2	17	Manipur	18
3	Assam	1212	18	Meghalaya	18
4	Bihar	715	19	Mizoram	3
5	Chandigarh	23	20	Nagaland	12
6	Chhattisgarh	1028	21	Orissa	875
7	Delhi	145	22	Punjab	2784
8	Gujarat	1924	23	Rajasthan	777
9	Haryana	650	24	Sikkim	18
10	Himachal Pradesh	171	25	Tamil Nadu	5186
11	Jammu & Kashmir	150	26	Tripura	55
12	Jharkhand	169	27	Uttar Pradesh	2216
13	Karnataka	1979	28	Uttrakhand	381
14	Kerala	1437	29	West Bengal	1600
15	Madhya Pradesh	754			

Table 3: Production of processed products in India

S.No.	Products	Production (,000 tones)
1.	Mango pulp	737.10
2.	Fruit juice	293.83
3.	Fruit nectars	125.70
4.	Dried fruits	84.8
5.	Preserves and candies	73.7
6.	Fruit Powder (flour)	15.4
7.	Mixed fruit pulps	3.3
8.	Juice concentrate	0.5
	Total	**1334.3**

Table 4: India's export of processed products

S.No.	Item	2016-17	
		Quantity (,000 MT)	Value (Crores)
1.	Dried and preserved vegetables	87.280	1088.55
2.	Mango Pulp	135.621	864.97
3.	Pickle & Chutney	180.820	942.72
4.	Alcoholic & Non-alcoholic beverages	231.332	2000.62
5.	Other processed fruits & Vegetables	351.83	3116.07
6.	Miscellaneous prepared products	283.265	2570.48
		1270.148	10583.41

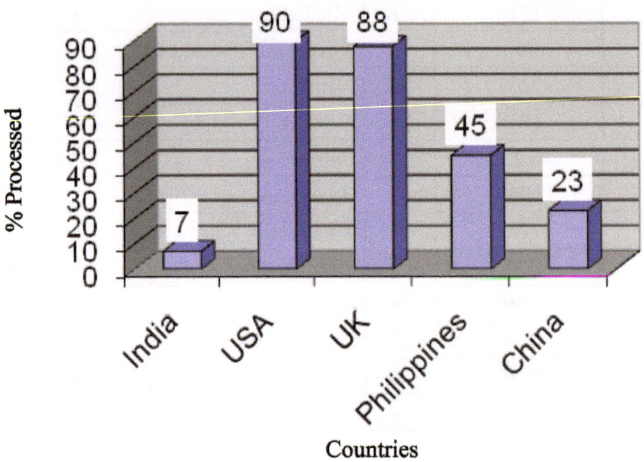

Fig. 2. Status of food processing in India in comparison to advanced countries

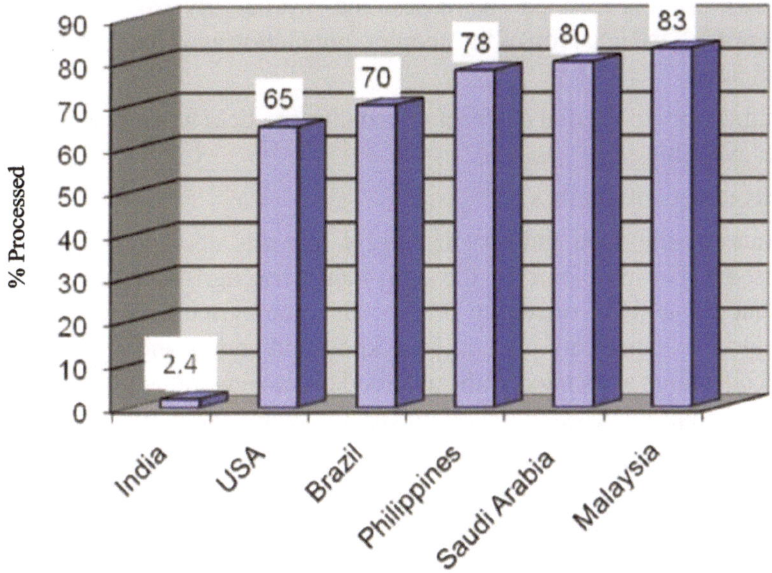

Fig. 3. Status of fruits and vegetable processing in India in comparision to advanced counties

Importance of Food Processing

Food processing industry is recognized as sun-rise industry. Food supply has to keep pace with the needs of the population. There is always a shortage of food in developing countries like India meet this, storage results in wastage due to inadequate facilities available for storage and preservation. It is therefore, imperative to expand facilities for the storage and preservation of food. Food processing is important because it helps in

1. Increasing the shelf life of food, thus increasing supply
2. Making the seasonal food available throughout the year and to such area where this
3. Saving time by reducing preparation time and energy
4. Stabilizing prices throughout the year.
5. To meet food requirement during natural calamity foods wars famines
6. Reduces wastage and losses.
7. Handles gluts.
8. Generate employment.
9. Stabilizes farm prices and income
10. Earn foreign exchange.

11. Preservation/processing increases the availability of foods, thus improving the nutrition of the peoples /population and checks nutritional deficiency
12. Add variety to the diet (to meet mixed diet practice) when seasonal food not available

Problems and Constraints

The fortune of the food industry is largely depends upon the caprices of the monsoon. The problems of the agro-industries, agri-businesses or any other economic activity related to or depends upon agriculture are therefore compounded. It is therefore imperative to know and understand the problems, risk and constraints associated with such food processing industries. The main constraints and problems of the food processing industry are given as under:

1. Food commodities are biological in nature and inherently subjected to variation in their physico-chemical composition and sensory characteristics as influenced by various factors.
2. Low productivity of the agricultural produce and non-availability of good quality food commodities for processing purposes.
3. Food commodities are mostly perishable or semi-perishable, leading to high degree of wastage of the produce.
4. Wide and violent fluctuations in the supplies, quality and price of the food as commodities may be location/region-specific or season specific.
5. Low level of developed technology and management inputs.
6. Lack of awareness about procedures and technologies to improve shelf life.
7. Inadequate infrastructure for storage, distribution and marketing.
8. Weak linkages in the supply chain
9. Lack of integration with upstream (food production) and downstream (marketing and export) agencies.
10. Very less value addition and processing

Prospects of Food Processing Industry

- Reduces wastage and losses
- Handles gluts
- Generate employment
- Stabilizes farm prices and income
- Checks nutritional deficiency
- Earn foreign exchange.

Emerging Technologies

Foods deteriorate in quality due to a wide range of reactions that may be physical, chemical or biological. The various forms of spoilage and food poisoning caused by micro-organisms are preventable to a large degree by a number of preservation techniques, most of which act by preventing or slowing microbial growth. These include freezing, chilling, drying, curing, conserving, vacuum packing, modified atmosphere packing, acidifying, fermenting, and adding preservatives. Thermal processes such as blanching, pasteurization or heat sterilization have long been employed in practice as economic, efficient, reliable and safe food preservation methods. However, in most cases thermal energy induces various chemical reactions, leading to quality deterioration in certain foods by producing undesirable changes in sensory and nutritional qualities. Thus, preservation technologies that prolong shelf-life without these detrimental effects are to be favoured. There are several new technologies of potential interest to the industry including high-pressure processing (HPP), pulsed electric fields (PEF), High intensity pulsed light, irradiation, ultrasound, extrusion cooking and modified atmosphere packaging (MAP). Among the novel technologies, the application of HPP, PEF and ultrasound to food processing and preservation has become of considerable interest in food research and development. These technologies are of specific interest to the food industry, because they not only provide attractive alternatives to conventional methods of thermal processing, which often produce undesirable changes in foods, affecting the balance between high quality and safety, but also offer opportunities for creating new ingredients and products because of the specific actions on biological materials and food constituents.

1. High Pressure Technology

High pressure processing (HPP) is a non-thermal process using water under very high hydrostatic pressure to produce packaged foods that are safer, longer lasting, more natural and better tasting. In addition, high pressure processing (sometimes also called high hydrostatic pressure processing or high pressure pasteurization) provides the food with new functional properties of ingredients such as hydrocolloids, proteins, etc. HPP is also called Pascalization or bridgmanization. It is a method of preserving and sterilizing food under very high pressure, leading to the inactivation of certain microorganisms and enzymes in the food. The technique was named after Blaise Pascal, a French scientist of the 17th century who studied the effect of pressure on fluids. During pascalization, more than 50,000 pounds per square inch (340,000,000 Pa) may be applied for around fifteen minutes, leading to the inactivation of yeast, mold, and bacteria. Pascalization is also known as bridgmanization named for physicist Percy Williams Bridgman.

It is the technique of the processing of the liquid food under high pressure. In general, pressure over 400 Mpa at 20-25°C is used to kill the vegetative microorganisms while a higher pressure of about 900 MPa combined with elevated temperature is needed to eliminated the spores. The HPP do not affect the quality of the food. It stabilizes the colour and maintain sensory characteristic along with microbial safety of food.

Applications of high pressure technology

- Preservation of fruit juices by inhibiting the growth of heat resistant bacterial spores.
- Cloud stabilization in orange juice.
- Enzyme activation and inactivation.
- Quality retention (especially flavour and colour)
- Modification of biopolymers, such as protein denaturation.

Advantages of HPP

1. Inactivate or kill food borne pathogens
2. Maintain nutrition
3. Maintain sensory properties
4. Inactivate spoilage organisms
5. Activate or inactivate enzymes
6. Inactivate some bacterial spores
7. Extend shelf life
8. Reduce the potential for foodborne illness
9. Pressure-shift freezing or thawing
10. Promote ripening of cheeses
11. Minimize oxidative browning

2. Pulse Electric Field

Pulse electric field (PEF) is one of the most promising emerging non-thermal processes. PEF processing of foods involves the application of short pulses (micro second to nano second) of high electric field intensity. The technique is similar to that of electroporation which is used successfully in biomedicine, cell biology and biotechnology. Based on this, PEF treatment is applied to liquid foods for inactivation of microorganisms and denaturation of enzymes. The exposure of cell membrane to a high-intensity electric field can be temporarily destabilized or irreversibly ruptured due to electroporation. The electrical damage may involve various mechanisms: critical transmembrane potential, visco-elastic properties, structural defects in the cell membrane, compression of the membrane and colloidal osmotic swelling. Among various emerging non-thermal technologies, high intensity pulsed electric fields (PEF) is one of the most important technique due to its short treatment times and reduced heating effects. Its capability to inactivate microorganisms and to enhance extraction processes at temperatures that do not cause any deleterious effect on flavor, color or nutrient value of foods, opens interesting possibilities for the food processing industry. PEF holds potential as a type of low temperature alternative pasteurization process. PEF technology enables inactivation of vegetative cells of bacteria and yeasts in various foods.

High intensity pulsed electric field (PEF) processing involves the application of short pulses (duration of micro- to milliseconds) of high electric field intensity (20 - 80 kV/cm) to foods placed between two electrodes. PEF treatment is conducted at ambient, sub-ambient, or slightly above ambient temperature for a very short time (less than 1 s), and thus energy loss due to heating of foods is minimized. In general, higher field strengths are used, up to about 35,000 V/cm for disinfection (destruction of bacteria, fungi, and other microbes). The lack of heat treatment makes PEF somewhat comparable to irradiation as a treatment and generally do not kill microbial spores or inactivate enzymes. So, the survival of spores and enzymes means that products should be refrigerated after passing through PEF processing in order to slow the action of the enzymes and prevent growth of microbial spores.

Thus, PEF means the processing of the food with the application of the high electric field strength (20-80kV per cm) with short pulses (1-20 μs) to samples placed between the electrodes. This operation takes place at low temperature of less than 50°C. It is generally used for pasteurization of the juices. The treatment with PEF also used to increase the yield of the juice during extraction.

Components of PEF Systems

The PEF processing system consists of 5 major components. These are:

1. High-voltage power supply
2. Energy storage capacitor
3. Treatment chamber
4. Pump to conduct food though the treatment chamber, a cooling device voltage, current, temperature measurement devices, and
5. Computer to control operations.

Advantages of PEF

1. The process is a gentle and waste-free technology.
2. It reduces detrimental changes in physical and sensory properties.
3. PEF process has potential for continuous application.
4. Process is instant and uniform for electrically conductive food.
5. PEF needs short treatment time.
6. It requires low energy.
7. Process help to retain colour, flavour and nutritive value of food to great extent
8. This technology offers great potential to the food industry because it can be applied to a wide range of products.
9. Process is effective to improve heat and mass transfers.

3. Extrusion Technology

Extrusion is a process that combines several unit operations including mixing, kneading, shearing, heating cutting, shaping and forming. It involves compressing and working of a material to form a semisolid mass under a variety of controlled conditions and then forcing it to pass through a restricted opening such as a shaped hole or slot at predetermined rate. It is a process of cooking moistened starchy and / or proteinaceous food material by combination of heat, pressure and mechanical shear. Single screw or twin screw extruder is used for extrusion cooking. It is a high temperature short time process. The act or process of extruding a product is called extrusion.

Extrusion involves compressing and washing raw material eg. Flours, starches, proteins, salt, sugar and other minor ingredients to form a semi solid mass under a variety of controlled conditions and then forcing it to pass through a restricted opening such as a shaped hole or slot at a predetermined rate. Heat is applied directly by steam injection or indirectly through a heated barrel or the conversion of mechanical energy. The final process temperature in the extruder can be as high as 200^{0}C but the residence time is relatively short i.e. 10-60 seconds. Thus extrusion cooking is also called high temperature short time (HTST) process.

Extrusion cooking is size enlargement process in which smaller granular foods or powdered particles are reformed into larger pieces. The main purpose of extrusion cooking is to the variety of foods in the diet, by producing a range of products with different shapes, textured, colours and flavours from basic ingredients. Extrusion cooking is high temperature short time (HTST) process which reduces microbial contamination and inactivates enzymes. However the main method of preservation of both hot and cold extruded food is by the low water activity of the product.

Extruders have many different size and shapes and allow a wide range of moisture content (10-40%), feed ingredients cooking temperature and residence times. The ability of extruders to vary the screw, barrel and die configurations for the cooking and shaping requirements of different foods products makes extrusion cooking one of the most versatile food processes. The application of extrusion cooking has been classified into (i) Semi finished products (ii) finished products. For semi finished products extrusion cooking offers economic advantage over traditional processes for producing pre-gelatinized cereal flours, potato starch, and cereal starches. It also include breakfast cereals, snack foods, textured food, dry and soft moist pet foods, beverage basis, pre cooked noodles, soup and gravy basis, fruit gums and chocolate.

Applications

Extrusion technology successfully replaced certain conventional processes of food processing and has a number of potential applications in making pasta, breakfast cereals, snacks (Directly expanded or "second generation" snacks), Texturised vegetable protein, Crisp bread, Precooked/instant flours, Co-extruded foods, Confectionery products, Animal feeds and pet foods, Rice bran stabilization.

Advantages of Extrusion

1. Reduction in raw material cost and offer saving of raw material upto 90%
2. Reduction in energy consumption and save 100% energy.
3. Rapid processing and so save time
4. Produce wide range of products
5. It has low process cost than other process.
6. Saving of labour is 14%.
7. Saving in capital investment is 44%
8. Extrusion process allow high production rate. That is it can produce snack food @ 315 kg h^{-1}, low density cereals @1200 kg/h^{-1} while dry expanded pet food @ 9000 kg/h^{-1}.
9. No process effluents are produced.
10. Produce protein based products TVP

11. Destroy inherent enzyme in soyabean lipase, Destroy urease, lipoxygenase in soya oil
12. Destroy anti-nutritional factors e.g. Trypsin inhibitor in soybeans

Limitations

1. Impart minor effect in colour and flavor
2. Fading of colour due to size expansion
3. Development of off flavor
4. Excessive heat reduce sugar metal ions.
5. Mallards reactions due to high temperate

4. Nanotechnology

The word "*nano'* meaning *'dwarf" in* Greek language refers to dimensions on the order of magnitude of 10^{-9}. Nanotechnology is nothing but manipulating of particles called nano-particles. These particles have dimensions in the scale of nanometer and used to design / build materials and devices. Nanotechnology has been provisionally defined as relating to materials, systems and processes which operate at a scale of 100 nanometers (nm) or less. Nanotechnology is an anticipated manufacturing technology that allows through, inexpensive control of the structure of matter by working them with atom. It will allow many things to be manufactured at low cost and with no pollution. The concept of nanotechnology was first given by Nobel Laureate physicist Richard P. Feynman in Southern California in 1959. He described vision of factories using nano-machines to build complex products, including additional nano-machines. This original vision for nanotechnology is sometimes termed Molecular Manufacturing or "MNT" (for molecular or molecular manufacturing- based nanotechnology). One nanometer (nm) is one thousandth of a micrometer (μm), one millionth of a millimeter (mm) and one billionth of a meter (m). *Nanomaterials have been defined as having one or more dimensions measuring 100 nm or less, or having at least one dimension at this scale which affects the materials behaviour and properties.*

Nanotechnology can make products cost effective (production will be carried out by self replicating nano-devices using small amount of material, energy, low capital, less labour and land), production being more efficient, decreased use of water and minimum wastage. However, there is a necessity for assessment of risk and means to control the same in order to reap its beneficial effects. Nanotechnology has the potential to revolutionize the scientific world by allowing scientists to manipulate matter at the atomic or molecular scale using physics, engineering, chemistry and biology. It enables researchers to understand the relationship between macroscopic properties and molecular structure in biological materials of plants and animal origin. Using nanotechnology, scientists are able to self assemble atoms into structures with highly controlled properties e.g. nanowires, self assembled molecules and particles, 3-D architecture etc. The concept of self- assembly with nanotechnology has the potential to impact diverse fields ranging from biology to materials science.

Nanotechnology can work from the top- down (which means reducing the size of the smallest structures to the nanoscale e.g. (photonics applications in nano-electronics and nano-engineering) or the bottom- up (which involves manipulating individual atoms and molecules into nanostructures and more closely resembles chemistry or biology).

Applications of Nanotechnology

- Single molecule detection to determine enzyme/ substrate interactions
- Nanocapsules for delivery of pesticides, fertilizers & other agrichemicals more efficiently
- Delivery of growth hormones in a controlled fashion
- Nanosensors for monitoring soil conditions and crop growth
- Nanochips for identity preservation and tracking
- Nansensors for detection of animal and plant pathogens
- Nanocapsules to deliver vaccines
- Nanoparticles to deliver DNA to plants (targeted genetic engineering)

5. Ohmic Heating

Ohmic heating, an innovative thermal process technology is currently utilized throughout Europe, Asia and North America. It is novel sterilization process in which heat is generated within a food by the passage of electric current, making it possible for particles to heat at the same or faster rate than fluids (an ideal ohmic process) whereas in practical scenario, heating rates depends on the electrical and thermal properties of both particle and liquid. This technique utilizes the principle that heat can be generated throughout a material by the passage of electric current, using the inherent electrical resistance of the material. Ohmic heating is defined as a process where primary alternating electric current is passed through a food or conducing solution (salt brine) with the purpose of heating them. It is also known as Joule heating, electric resistance heating, direct electrical resistance heating, electro heating and electro conductive heating. In ohmic heating, the presence of the electrode results in heating of the foods. The heat transfer problems have now been overcome with the development of ohmic heating which operates by the direct passage of electric current through the food product.

Further these thermal, chemical and mechanical effect on microorganisms are achieved through ohmic heating process by electric heating, electrohydrualic shock, electroporation and electroconductive heating as illustrated in Table 1.

Table 1. Mode of action of preservation in ohmic heating.

S.No.	Mode of action	Result	Microorganism involved
1.	Electric heating	Generation of heat causes bacterial death. This type of pasteurization is called "electropure process".	*E. coli* *M. tuberculosis*
2.	Electrohydraulic shock	Heat shock which causes mutations formation of free radicals, membrane damage	*E. coli* *Staphylococcus aureus*
3.	Electroporation	Disruption of cell membrane	-
4.	Electro-conductive heating	2-3°C / sec temperature rise at centre of particulate and inactivation of peroxidase.	-

Advantages of ohmic heating

- Rapid and uniform heating of the particulate food.
- Continuous production.
- Higher and rapid sterilization at low cook values.
- Higher quality products.
- Increased nutrient and flavour retention with particulate integrity
- Processing of high solid content as compared to conventional methods.

Application of ohmic heating

Ohmic heating can be applicable to preserve pasta sauce with vegetables, tomato sauce, sliced kiwi fruit in sugar syrup, ham etc. It can be used to process meat chunks, black currants, whole strawberries, blackberries, diced and sliced vegetables, baked beans and mushrooms including liquid food products which are difficult to heat by conventional methods like fruit juices. Ohmic heating can be used for heating of high acid food products such as tomato based sauces prior to hot-filling. Broccoli florets in cheese sauce, shrimp, okra, tomatoes, corn pea pods, bamboo shoots, carrots and red peppers in a savory sauce etc. have successfully been processed in ohmic heater.

6. Microwave Heating

These are electromagnetic waves between infrared and radio waves. Microwave is comparatively short non ionizing electromagnetic wave having one millimeter to one meter in wavelength and has frequency between 300 MHz and 300 GHz. These waves are used to cook or heat food in a microwave oven. Magnetron results in production of the microwaves. Magnetron is a vacuum tube in which the flow of electrons is controlled by an applied magnetic field to generate power at microwave frequencies. A magnetron is a high power microwave oscillator in which the potential energy of an electron cloud near the cathode is converted into r.f. energy in a series of cavity resonators.

Applications of microwave heating

- *Drying at Atmospheric pressure:* The drying of pasta is an established application comprising three stages involving microwaves and hot air in various combinations, to give improved sanitation and better control as well as quality. Other examples include the drying of onions, parsnips, snack foods (with subsequent expansion as described above in puffing of pellets), fabrics, leather, ceramic cores and moulds and ceramic wares.
- *Vacuum drying:* Some materials are heat sensitive and cannot be dried at atmospheric pressure. It is necessary to reduce the pressure to reduce the boiling point and effect drying at a reduced temperature. A modest vacuum around 100-200 mm Hg is necessary where the formation of a microwave plasma or arc can still be avoided. Notable examples are the drying of fruit juices, beverages, drugs and pharmaceutical pellets.
- *Heating and cooking:* Many foodstuffs have been cooked by microwaves for various stages of processing. Examples include bacon cooking in a combination system, meat coagulation to upgrade scrap and doughnut cooking for frying.
- *Pasteurisation and sterilization:* Food products, such as bread, precooked foods and animal feedstuffs have been processed using microwaves for pasteurisation or sterilisation or simply to improve their digestibility. Specific examples include the sterilisation of bonemeal and the processing of barley to achieve starch to gelatine conversion. Food pasteurization of sealed packs under pressure can be effected by microwave energy; however, as with most pasteurization processes the product after treatment needs rapid cooling to avoid infestation.
- *Baking:* Microwave heating also plays an important role in bakery industry. It is used in making biscuits, cakes, breads etc.

- *Microwave* heating is used to preheat the snack foods before serving as it is a short time process.
- Microwave heating is used for blanching of vegetables prior to further processing.
- Pre-treatment with microwave heating help in increase juice yield.
- Microwave heating help to inactivate anti-nutritional factors.
- Microwaves heating is used to thaw the frozen foods.

Advantages of microwave heating

- Rapid heat transfer.
- Saves times of processing or heating.
- Microwave oven consumes very less electricity and is easy to use.
- Easy to heat the foods
- Volumetric and selective heating.
- Retains nutrients and vitamins in the food
- Easy to clean and maintain.
- Compactness equipment results speed of switching on and off.
- The heating is pollution-free environment as there are no products of combustion. Microwave leakage can certainly be kept well below government recommended levels.
- Blanching with microwaves can check leaching losses of nutrients.

Disadvantages

- Microwave heating do not result formation of crust in the bread surface.
- Moisture is needed for heating. Dried products cannot be heated in microwave.

Precautions

- Closed containers and raw eggs can burst when heated in a microwave oven due to the stress build-up of steam.
- Liquids if overheated may result in violent burst of water and vapor resulting in liquid and steam burns.
- Use cookware meant for microwave cooking. Do not use metal or plastic cookware to cook in the microwave oven.
- Do not forget to wear gloves while taking out the food from the oven.
- If your microwave oven is not working, then do not try to repair it by yourself.

- If you think that something is wrong with your microwave oven like spark or fire, put off the main switch with the use of the wooden stick or any other insulated object. microwave oven.

7. Ultrasonication

The application of ultrasound waves in the preservation of the food is known as ultrasonication. Piezoelectric effect is observed in ultrasonication. Ultrasound comprises vibrations similar to sound waves, but of a frequency too high (18kHz-500MHz) to be heard by the human ear. Such sound waves in combination with pressure and heat have more effect than the individual. Such combinations include thermosonication (heat + ultrasound), manosonication (pressure + ultrasound) and manothermosonication (pressure + heat + ultrasound). These combinations are more effective against microorganisms and enzymes. The combined use of heat and ultrasound, or heat and ultrasound under pressure, markedly increases the lethality of heat treatments and consequent reductions in time and/or temperature of heat processes. A combination of low pressure (0.3 MPa), heat treatment and ultrasonic wave treatment for preservation of the food is called manothermo sonication. Sonicator is used for generation of the ultrasounds. It is an instrument using sound waves for the preservation of the food. Sometime, it is also called as sonication. Ultrasonication at high enough intensities has long been known to inactivate vegetative bacteria and to reduce the heat resistance of spores; the effect is amplified by increasing the temperature. Ultra-sonication accelerates both physical and chemical reactions, enhancing surface chemistry, and causing violent particle motion, which generates high velocity inter-particle collisions. Ultrasonic techniques are finding increasing use in the food industry for both the analysis and modification of foods. Low-intensity ultrasound provides information about physico-chemical properties, while high-intensity ultrasound is used to alter either physical or chemical properties of foods, e.g. to generate emulsions, disrupt cells, promote chemical reactions, inhibit enzymes, tenderise meat and modify crystallisation processes. The use of ultrasonics alone for microbial inactivation of foods appears ineffective, hence use of ultrasounds in combination with other anti-microbial methods increase its effectiveness against microbial and enzyme inactivation. Such combinations include thermosonication (heat + ultrasound), manosonication (pressure + ultrasound), manothermosonication (pressure + heat + ultrasound) and osmo-sonication (osmosis + ultrasound). These combinations were found to be effective against microorganisms and enzymes. The combined use of heat and ultrasound, or heat and ultrasound under pressure, markedly increases the lethality of heat treatments and consequent reductions in time and/or temperature of heat processes. Combined heat and ultrasound treatments have been reported to lower maximum processing temperatures by 25-50 % with less reduction vitamin C and less colour change.

The principles behind the ultrasonication is formation of cavitation (vacuum phase and implosion plase). The process starts with the formation of very small entities called 'microbubbles". These bubbles enlarges as much as possible. During compression lot of pressure (500 atm) is applied on these bubbles. When these bubbles cannot withstand any more pressure, they explode with the release of high energy and the temperature may rise upto 5000°C.

Benefits

- Ultrasonic treatments are very effective to inactivate enzyme and bacteria.
- Less time consuming.
- Beneficial in preparation of biomolecules by lysing of bacteria
- Very less maintenance is required as the probe is shelf cleaning.

Drawbacks

- This technique is not effective against bacterial spores.
- Sample heat up is very common and thus constant temperature cannot be maintained.

Suggested Readings

Bakshi AK, Joshi VK, Vaidya D and Sharma S 2013. Food Processing and Preservation. Jagminder Book Agency, New Delhi, pp 415-425.

Bawa AS, Raju PS and Chauhan OP 2013. Food Science. New India Publishing Agency, New Delhi, pp. 398.

Chaudhary M, Pandey M C, Radhakrishna K and Bawa A S. 2006. Nanotechnology: Applications in food industry. Bev. and Food World. 36: 60-62

Dev Raj 2011. Food Science and Technology: Glossary of preeminence. New India Publishing Agency, New Delhi, pp 388.

Dev Raj 2012. Post Harvest Technology and Engineering: An Illustrated Guide. New India Publishing Agency, New Delhi, pp 302.

Hanlon, JF 1984. Emerging trend in food packaging. Handbook of package engineering. 2nd ed. Mcgraw Hill Book Co. New York.

Holdsworth, S.D. 1983. The Preservation of Fruits and Vegetable Food Products, Macmillan Press, London, UK.

Hui, Y.H., Ghanzala, S., Graham, D.E., Murrell, K.D. and Nip, W.K. 2004. Handbook of Vegetable Preservation and Processing. Marcel Dekker, Inc.

Joseph T, Morrison M. 2006. Nanotechnology in Agriculture and Food. Nanoforum Report. available at: http://www.nanoforum.org/ dateien /temp/ nanotechnology %20in% 20agricult

Kakade N 2003. Nanotechnology: a new challenge. Electronics For You. 35: 31-36

Lee, C.Y. 1989. Green peas, Quality and Preservation of Vegetables, (N.A. Michael Eskin, ed.), CRC Press, Boca Raton, FL, USA. pp. 159.

Opara L. 2004. Emerging technological innovation triad for agriculture in the 21st century. Part 1. Prospects and impacts of nanotechnology in agriculture. Ag Engineering Internat: CIGR J Ag Engineering Internat Vol 6.

Potter, N.N. and Hotchkiss, J.H. 1995. Food Science 5th Edition. Chapman & Hall, New York USA

Rahman, M.S. 1999. Handbook of Food Preservation, Marcel Dekker, NY, USA.

Raj D, Sharma R and Joshi VK 2011. Quality control for value addition in food processing. New India Publishing Agency, New Delhi, pp 324.

Roco M C, Williams R S, Alivartos, P. 1999. Nanotechnology Research Directions: IWGN Workshop Report, Kluwer Academic Publishers: Norwell, MA,

Sharma SK 2010. Post harvest management and processing of fruits and vegetables. New India Publishing Agency, New Delhi, pp 390.

2

Trends in Food Safety Management

S.K. Sharma and Deepa Saini

Department of Food Science and Technology, College of Agriculture, GB Pant University of Agriculture and Technology, Pantnagar, Uttarakhand

Food Safety

The term "safe food" has different standards to different consumers rather than just being the food handled and prepared through proper cleaning, washing and cooking. Safe food means food prepared on clean and sanitized surfaces with equipment, utensils and dishes that also are cleaned and sanitized. Nowadays consumers are extra conscious about the safety and quality of food and for that reason food safety issues have changed, so have society's methods for making food as safe as possible. Accordingly, "food safety" means assurance that food is acceptable for human consumption according to its intended use (FSSAI 2017).

Since time immemorial human beings, consciously or unconsciously, have been concerned about the safety of food. Through their knowledge, they selected wild plants, animals for survival, skillfully avoiding, as far as possible, poisonous species. In global food market now consumer safety is the top most priority. To fulfill this is a hard task because food is very complex commodity and has to face number of biological, chemical and physical hazards. Due to contamination of food with these hazards the world population is suffering from number of short term, long term illnesses or deaths. Almost 70 % of the illnesses now prevailing in the society are due to food, water, air borne microbial infections and intoxications, heavy metal, pesticide residues, veterinary drug and other chemical residues toxicities. The infections are causing sporadic, epidemic and pandemic out breaks throughout the world. We still remember severe pandemic in Europe by swine flu / influenza, which killed more than 50, million people during 1918-20 and in 2009-10 WHO has declared swine flu as pandemic disease affecting peoples of around 203 countries. Number of such biological agents is active in and around food from water, environment and soil. In addition to this safety issues of transgenic foods are also questioned in international market.

Currently around 85,000 chemical agents are produced and used in industrialized countries out of this less than 45% have undergone basic toxicological tests and many of them are causing severe toxicities and casualties leading to gastroenteritis, congenital malformations, asthma, cancer and neurological and behavioral disorders. Majority of these chemicals are found in essential commodities such as water, air, soil, and foods without which we can't survive i.e., why henceforth restriction in terms of sanitary and phytosanitary measures both for import and export of agriculture commodities are imposed and accordingly the residues of pesticides and chemicals are to be reduced to internationally acceptable levels.

Hazards

Since, procurement of raw material from different sources as plants and animals, there may be incidence of hazards (physical, chemical and biological), or hazards may introduced in the product during the processing such as, chipped metal and lubricants from machinery, grimy processing vessels, unhygienic handling of raw materials, dirty premises and there are many more reasons for hazards. But, there is need to clearly define the hazards and what is the critical limit beyond which it is considered to be hazardous (FAO 2005). ***Hazard is a biological, chemical or physical agent that is reasonably likely to cause illness or injury in the absence of its control.*** The hazard is linked with toxicity i.e. the inherent capacity to come harm. Hazards can be divided into following groups.

- Biological / microbiological hazards
- Chemical hazards
- Physical hazards

In recent years, some more categories have been added to these types, but they require debate and substantiation as per definition of hazards. Some of such categories of hazards are:

- Allergens
- Nutritional hazards
- Biotechnology related hazards / novel foods

1. Biological hazards

It includes fungi, bacteria, viruses, parasites, insects and pests. These are responsible for foodborne hazards in any type of food establishment. These cause most foodborne illnesses and are the primary target of a food safety program. These agents require food, moisture, and warm climate to multiply.

Foodborne infection: When a living, disease-causing microorganism is ingested along with a food, it can cause a foodborne infection. After ingestion, the organism burrows into the lining of the victim's digestive tract and begins to grow in number. This can lead to the common symptoms of foodbrone illness such as diarrhea. Bacteria, virus, and parasites are examples of microorganisms that can cause infection.

Food intoxication: It is caused when a living organism multiplies in or on a food and produces a chemical waste or toxin. If the food containing the toxin is ingested the toxin causes an illness. The common examples of food intoxications are *Clostridium botulinum* and *Staphylococcus aureus*. Intoxication may also occur when an individual consumes food that contains manmade chemicals such as cleaning agents or pesticides. This is also sometimes used synonymously with *food poisoning*. The hazardous microbes are categorized degree of severity which is based upon pathogenicity, number of consumers affected, infective dose, case fatality ratio, pathological effects on the host, persistence in the body of the host, duration of illness, cross-infection, etc.

In addition to disease caused by direct bacterial infection, some foodborne illnesses are caused by exotoxins which are excreted by the cell as the bacterium grows. Exotoxins can produce illness even when the microbes that produced them have been killed. Symptoms typically appear after 1–6 hours depending on the amount of toxin ingested. *Clostridium botulinum*, *Clostridium perfringens*, *Staphylococcus aureus*, *Bacillus cereus*. Sometimes, infection is caused by viruses also but viral infections make up perhaps one third of cases of food poisoning in developed countries. In the US, more than 50 % of cases are viral and noroviruses are the most common food borne illness, causing 57 % of outbreaks in 2004. Foodborne viral infection is usually of intermediate (1–3 days) incubation period, causing illnesses which are self-limited in otherwise healthy individuals, and are similar to the bacterial forms.

Principal Biological Hazards

Bacteria		
• *Bacillus cereus* • *Campylobacter jejuni* • *Clostridium botulinum* • *Clostridium perfringens* • *Escherichia coli* 0157:H7	• *Escherichia coli* 0104:H4 • *Listeria monocytogenes* • *Salmonella* spp. • *Shigella* spp. • *Staphylococcus aureus*	• *Vibrio cholerae* • *Vibrio parahaemolyticus* • *Vibrio vulnificus* • *Yersinia enterocolitica* • *Cronobacter sakazakii*
Viruses		
• Bacteriophage • Hepatitis A virus • Norovirus	• Norwalk virus • Rota virus	• Enteric Virus (other than Hepatitis A and Noroviruses)
Parasites		
• *Cryptosporidium parvum* • *Giardia duodenalis* or *intestinalis*	• *Taenia* spp. • *Toxoplasma gondii* • *Trichinella spiralis*	• *Entamoeba histolytica* • *Entamoeba coli*
Others - Prions		

2. Chemical Hazards

These types of hazards can be classified as naturally occurring chemicals and added chemicals. Naturally occurring substances are the toxic substances present in the plants, they are inherent toxins and acquired toxins. Inherent toxins are protease inhibitors, hemagglutinins, goitrogens, cyanogenic glycosides, lathyrogens, saponins, allergens, vasoactive pressure amines, vicine and convicine, carcinogens, flatulence causing factors, specific anti-nutrients such as anti-vitamins etc. While, acquired toxins are the microbial toxin produced in the food by microorganisms such as *B. cereus*, *S. aureus* and *C. botulinum*. For example, mycotoxins that is toxin produced by fungus. Some of the mycotoxins which are most important from the food safety point of view aflatoxins, ochratoxin A, fumonisins,citrinin, patulin, ergot alkaloid, trichothecenes, alternariol and zearalenone.

Endotoxins: Endotoxin is the name given to a group of heat stabile lipopolysaccharide molecules present in the cell walls of gram-negative bacteria that have a certain characteristic toxic effect. The lipid portion of each molecule is responsible for its toxicity and can vary between bacterial species and even from cell to cell. When inhaled, endotoxin creates an inflammatory response in humans that may result in fever, malaise, alterations in white blood cell counts, headache, respiratory distress and even death. It is common to the environment due to the ubiquitous nature of Gram-negative bacteria. Exposure to elevated levels of endotoxin primarily occurs through exposure to aerosols from specific reservoirs such as cotton mills, wastewater treatment facilities, air washers, humidifiers and any other occupational settings where Gram-negative bacteria can flourish.

Some other types of contamination caused by chemicals are due to heavy metals, pesticides, residues from veterinary drugs, additives and chemicals developed during processing and genetically modified foods. Heavy metal contamination is a general term given to describe a condition in people who have abnormally high levels of toxic metals in the body. Common examples are ***mercury, lead, cadmium and arsenic***. This contamination can be very real, detrimental to health and deadly. Heavy metals are subtle, silent, stalking killers. They enter the body with our food, water, air we breathe and by skin contact. They slowly accumulate in the kidneys, liver, pancreas, bones, central nervous system and brain where they degrade health without being noticed or diagnosed. Heavy metals can cause cancer, high blood pressure, heart disease and mental retardation. Pesticide residues are also hazardous to human health. WHO-FAO set rules for safety standards for member counties of WTO to domestic and imported food products for safeguard the consumer health from unreasonable health risk posed by pesticide residues. Food safety authorities set standards for additives also, however various chemicals developed during the processing of food at high temperature. The roasted foods are reported to be containing a chemical known as *acrylamide* formed by reaction between protein & carbohydrate. It is present in various fried & baked foods in the range of 10 to 4000 ppb. It is known to involved in carcinogenesis, causes neurological damage, genetic mutation and impaired fertility. The present daily intake of acryl amide in US, Europe and other industrialized countries could be as high as 18 ug./ Kg. of body weight while current safe level for this toxic chemical in water is fixed at 0.2 mg./kg. of body weight. The Nestle Lab. reported that carbohydrate (dicarbonyls) also can react with amino acids like glycine, methionine & cysteine beside asparagine, to produce acrylamide under simulated conditions. Food safety issues are concerned with genetically modified foods as well since some people worry that live GMF's could transfer antibiotic resistant genes to people and it could limit the number of antibiotics in medical treatment. Several crops are genetically modified for the improved nutritional quality by altering DNA or insertion of gene from other plant or microorganism. It was observed that sometimes there is risk of allergencity associated due to these alterations. Because of aforesaid reasons GMF's does not allowed in some countries as food standards and regulations approval.

Principal Chemical Hazards

Mycotoxins		
• Aflatoxin • Ochratoxin A	• Fumonisin • Patulin	• Deoxynivalenol (Vomitoxin)
Natural toxins		
• Glycoalkaloids • Hypoglycin • Cyanogenic glycoside • Lectin	• Amanitins • Gyromitrin • Muscarine • Phallotoxins	• Furocoumarins • Oxalic acid • Goitrogens
Marine toxins		
• Saxitoxin • Domoic acid	• Ciguatoxin • Okadaic acid	• Tetrodotoxin • Brevetoxin
Environmental contaminants		
• Lead • Arsenic	• Bromates • Dioxins • Cadmium	• Furans • Mercury • Polychlorinated biphenols (PCBs)
Food additives		
• Food colours (natural and synthetic) • pH adjusting agents • preservatives	• bleaching agents • food enzymes • glazing and polishing agents	• emulsifiers • gelling agents
Processing induced chemicals		
• acrylamide	• ethyl carbamate	• furan
Pesticides		
• Azoxystrobin • Captan	• Clethodim • Thiocarbamate	
Veterinary drugs		

3. Physical hazards in the agri-food chain

Numbers of physical hazards like stones, metal pieces, hairs, pieces of pastics etc are involved in food which causes harm to health. Illness and injury can result from hard foreign objects in food. These physical hazards can result from contamination and/or poor procedures at many points in the food chain from harvest to consumer, including those within the food establishment. These hazards may cause cuts, bleeding, infection, choking, broken teeth, to the consumer.

4. Allergens

These substances are at times categorized separately as hazard or sometimes under chemical hazard, which is mystifying. Cereals (rice, wheat, barley, etc.), peanuts, peas, lentils, soya beans, strawberries, bananas, mangoes, pineapples, sesame, poppy and caraway seeds, tea, chocolate, coffee, yeasts, alcoholic beverages, honey and other foods likely to contain pollen restrain allergens, for example, caffeine, cyanide, favism, goitrogens, haemagglutinins, oxalic acid, nitrates and nitrites, pyrrolizidine alkaloids, solanine, etc.

Risk

Risk is the possible and irrepressible loss of valuable things. In case of food production, it is foremost requirement to reduce the chances of risk by limiting the introduction of any kind of hazard from any source and practice. So, it is essential to analyze the risk consisting risk assessment, risk management and risk communication.

Risk assessment is a scientifically based process consisting of the following steps:

i) Hazard Identification

ii) Hazard Characterization

iii) Exposure Assessment; and

iv) Risk Characterization

Risk management is the process of weighing policy alternatives in consultation with all interested parties, considering risk assessment and other factors relevant for :

a) Health protection of consumers

b) Promotion of fair trade practices

c) Selecting appropriate prevention and control options.

Risk communication is the interactive exchange of information and opinions throughout the risk analysis process concerning risk, risk-related factors and risk perceptions, among risk assessors, risk managers, consumers, industry, the academic community and other interested parties, including the explanation of risk assessment findings and the basis of risk management decisions.

Food safety management system

It is a system that describes the ability of an organization and what are the essentials they must do to control food safety hazards and ensure that food is safe for consumption. Comparatively to past period among the industrial groups, food industry is one of the fastest and biggest growing industry. With the higher expectations of consumers, their safety and well being should be a prime objective of an organization representing this industry. To maintain these requirements Food Safety Management Systems is the excellent tool and to ensure the safe production of food, it helps the firm to gain the trust of its consumers or even serve them efficiently. Implementation of prerequisites as good practices assures the quality product like, good agricultural practices (GAPs), good handling practices (GHPs) and good manufacturing practices (GMPs).

GAPs: GAP, as defined by FAO, are a "a set of principles to be applied to the production and post-production processes in the field and farm that create safe and healthy food and non-food agricultural products, taking into account economic, social and environmental sustainability". Both importing countries and domestic buyers, in particular organized retailers, require manufacturers to apply GAP as a prerequisite for procurement to assure the quality and safety of their products. Furthermore, the implementation of GAP also contributes to promoting sustainable agriculture and contributes to the achievement of environmental and social development goals at national and international level. It has been shown that the implementation of GAPs promotes the optimal use of resources such as pesticides, fertilizers and water, as well as an environmental friendly agriculture. Its social dimension would be to protect the health of agricultural workers from the abuse of chemicals and pesticides.

GHPs: Good handling practices confirms the minimum sanitary and hygienic practices for food processing by food industry, hotels, restaurants to assure that food is safe and apt for human consumption.

GMPs: GMPs umbrella address only the basic sanitary requirements for food processing, handling, and storage. GMPs are essential for the manufacture and distribution of foods that are safe from microbiological, chemical, and physical hazards. It is imperative that the food industry manage a comprehensive program that evaluates, identifies, and controls potential hazards at every step in the production, development, and manufacturing environment.

HACCP (Hazard Analysis and Critical Control Points)

HACCP is a systematic preventive approach to food safety and pharmaceutical safety that identifies physical, allergenic, chemical, and biological hazards in production processes that can cause the finished product to be unsafe, and designs measurements to reduce these risks to a safe level.

The HACCP seven principles

Principle 1: Conduct a hazard analysis. Plans determine the food safety hazards and identify the preventive measures the plan can apply to control these hazards. A food safety hazard is any biological, chemical, or physical property that may cause a food to be unsafe for human consumption.

Principle 2: Identify critical control points. A critical control point (CCP) is a point, step, or procedure in a food manufacturing process at which control can be applied and, as a result, a food safety hazard can be prevented, eliminated, or reduced to an acceptable level.

Principle 3: Establish critical limits for each critical control point. A critical limit is the maximum or minimum value to which a physical, biological, or chemical hazard must be controlled at a critical control point to prevent, eliminate, or reduce to an acceptable level.

Principle 4: Establish critical control point monitoring requirements. Monitoring activities are necessary to ensure that the process is under control at each critical control point. In the United States, the FSIS is requiring that each monitoring procedure and its frequency be listed in the HACCP plan.

Principle 5: Establish corrective actions. These are actions to be taken when monitoring indicates a deviation from an established critical limit. The final rule requires a plant's HACCP plan to identify the corrective actions to be taken if a critical limit is not met. Corrective actions are intended to ensure that no product injurious to health or otherwise adulterated as a result of the deviation enters commerce.

Principle 6: Establish procedures for ensuring the HACCP system is working as intended. Validation ensures that the plants do what they were designed to do; that is, they are successful in ensuring the production of a safe product. Plants will be required to validate their own HACCP plans. FSIS will not approve HACCP plans in advance, but will review them for conformance with the final rule. Verification ensures the HACCP plan is adequate, that is, working as intended. Verification procedures may include such activities as review of HACCP plans, CCP records, critical limits and microbial sampling and analysis. FSIS is requiring that the HACCP plan include verification tasks to be performed by plant personnel. Verification tasks would also be performed by FSIS inspectors. Both FSIS and industry will undertake microbial testing as one of several verification activities. Verification also includes 'validation' – the process of finding evidence for the accuracy of the HACCP system (e.g. scientific evidence for critical limitations).

Principle 7: Establish record keeping procedures. The HACCP regulation requires that all plants maintain certain documents, including its hazard analysis and written HACCP plan, and records documenting the monitoring of critical control points, critical limits, verification activities, and the handling of processing deviations.

HACCP is internationally recognized as a logical tool for adapting traditional inspection methods to a modern, science-based, food safety system. Based on risk-assessment, HACCP plans allow both industry and government to allocate their resources efficiently in establishing and auditing safe food production practices. In 1994, the organization of *International HACCP Alliance* was established initially for the US meat and poultry industries to assist them with implementing HACCP and now its membership has been spread over other professional/industrial areas.

Recent challenges in food safety

Sometimes the foods we love and count on for good health are contaminated with germs that cause sickness and can even be deadly. More progress is needed to protect people and reduce foodborne illness. New challenges to food safety will continue to emerge, largely because of:

- **Antibiotic resistance**: It compromises our ability to treat infections and is a serious threat to public health. *Salmonella* and *Campylobacter*, are commonly transmitted through food, cause an estimated 410,000 antibiotic-resistant infections in the United States each year. Antibiotic resistance is the ability of microbes (bacteria, viruses, fungi and parasites) to resist the effects of drugs. Changes in our food production and supply, including more imported foods.
- **Development of new strains**: New strains develop due to increase of multidrug resistance in pathogenic, opportunistic, and spoilage bacteria, which can reach humans through the food chain. In recent years, a new strain of *S. aureus*, livestock-associated methicillin-resistant *Staphylococcus aureus* (LA-MRSA), has been recognized as a novel pathogen that has become a rapidly emerging cause of human infections and that may lead to the with increased pathogenicity.
- **Variations in food safety standards**: Food safety standard being not harmonized in the all countries, there is a lot of variation in food handling and compliance with food standards.
- **Allergens**: Different consumers are allergic to different foods or ingredients. Some foods, such as nuts, milk, eggs, fish, crustacean shellfish, tree nuts, peanuts, wheat or soybeans, can cause allergic reactions in people with food allergies. This is a great risk because in this case consumers have to be alert regarding ingredients before consumption otherwise it may cause health problem.
- **Changes in consumer need and habits**: Due to growing urbanization, and altering consumer preferences and habits, there is an increase in the number of people preferring to buy and eat food prepared in public places.
- **Changing dining trends:** Display of FSMS, Certifications of food handlers and Hygiene ratings for hotels and restaurants. This is being

mad mandatory in India especially for online food suppliers.

- **Traceability is key to the experience:** Consumers want traceability of what they eat. This is driving a need for greater auditing of supply chains to ensure that food can be monitored from field to fork. As we move towards 2020, food businesses that can vouch for the origins of their food will have a major competitive advantage.
- **Food fraud:** Transparency throughout supply chain is strongly needed to create an environment of trust. Industry, today faces a challenge of holding the trust of customers. Self compliance and declarations, audit reports etc. help us here.
- **Digital Food Safety: At present food manufacturers and handlers** rely on paper-based checks and records that are inefficient, time-consuming, prone to human error and easy to falsify. Automated systems for monitoring, sensors for everything and digital checklists will replace paper in future. Even robots shall be used for monitoring and control of every aspect of food handling. Therefore digitization is the key for future food safety and compliance management, for better trust and facilitation of consumers as well as regulators.

Conclusion

Food safety is a dynamic process and the needs change with a large number of factors as discussed in this chapter. The key to survive in the present time is continuous adoption of newer solutions, self compliance, digitization and transparency of functioning. Food safety is not a choice for manufacturer, it is a moral, professional and business responsibility.

References

Anonymous 2019. Food Safety Risk Analysis. https://www1.agric.gov.ab.ca/$Department/deptdocs.nsf/all/afs12301/$FILE/appendix_d_risk_analysis.pdf.

Bennett PM (2008). "Plasmid encoded antibiotic resistance: acquisition and transfer of antibiotic resistance genes in bacteria," British Journal of Pharmacology, vol. 153, supplement 1, pp. S347–S357, 2008.

CDC 2013. Antibiotic Resistance Threats in the United States, 2013. Atlanta, GA: U.S Department of Health and Human Services, CDC. 2013; p.7: 36-7.

FAO 2005. Food Safety Risk Analysis PART I - An Overview and Framework Manual. Food and agricultural Organization, Rome.

FAO 2016. A scheme and Training Manual on Good Agricultural Practices for Fruits and Vegetables. Vol 1 – The scheme- standard and implementation infrastructure. Food and Agricultural Organization, Regional Office, Bangkok. http://www.fao.org/3/a-i6677e.pdf

FAO 2019a. Assuring Food Safety and Quality: Guidelines for Strengthening National Food Control Systems. FAO Food and Nutrition Paper 76. http://www.fao.org/3/a-y8705e.pdf

FAO 2019b. Risk analysis and food: The experts' view. http://www.fao.org/3/v9723t/v9723t08.htm

FSSAI 2017. Basic Food Safety Manual for Food Manufacturers, Processors and Packers. Food Safety and Standards Authority of India, New Delhi. www.fssai.gov.in

Kasbekar Arundhati 2018. Top food safety challenges of 2018 in Europe, the US and beyond. https://www.newfoodmagazine.com/article/64715/food-safety-2018/

Larson RL, Harper AL, Hanson BM et al., 2011. "Methicillin-resistant Staphylococcus aureus in pork production shower facilities," Applied and Environmental Microbiology, 77(2): 696–698

Niederhäusern S, Sabia C, Messi P, Guerrieri E, Manicardi G, and Bondi M., 2004. "Glycopeptide-resistance transferability from vancomycin-resistant enterococci of human and animal source to Listeria spp," Letters in Applied Microbiology, 39(6): 483-489.

3

Post Harvest Treatments for Fresh Fruits and Vegetables

J.M. Mayani and Dev Raj

Department of Post Harvest Technology, ASPEE College of Horticulture Navsari Agricultural University, Navsari, Gujarat-396450, India

Introduction

Many post-harvest treatments are applied to horticultural crops, either to maintain the quality (taste, colour, flavour, texture) or improve the visual appeal. Most important of these treatments are temperature management including the cold chain where the temperature of the crop is reduced rapidly and stabilize temperature after harvesting. Exposing the crop to high or low temperature and application of chemicals after harvest helps in managing/ prevent pest and diseases and sprouting occurrence respectively.

Harvested produce must be handled with care at every stage to avoid the mechanical damage and subsequent fungal/bacterial infection. Adopting appropriate post-harvest handling operation will minimize the all ill effects of post-harvest (Fig. 1).

Harvesting of Produce	→	Pre-cooling	→	Sorting/cleaning/trimming
				↓
Sizing/ grading	←	Chemical treatment/curing/waxing	←	Washing
↓				
Packing	→	Transportation	→	Wholesaler
				↓
Retailer	←	Transportation	←	Resorting, Resizing and Repackaging
↓				
Consumer				

Fig. 1. Flow chart for post-harvest handling operations of horticultural produce Postharvest treatments on Horticultural Produces

Combine machine which handle many operations such as washing, waxing and drying, sizing, grading and packing are being employed in pack house.

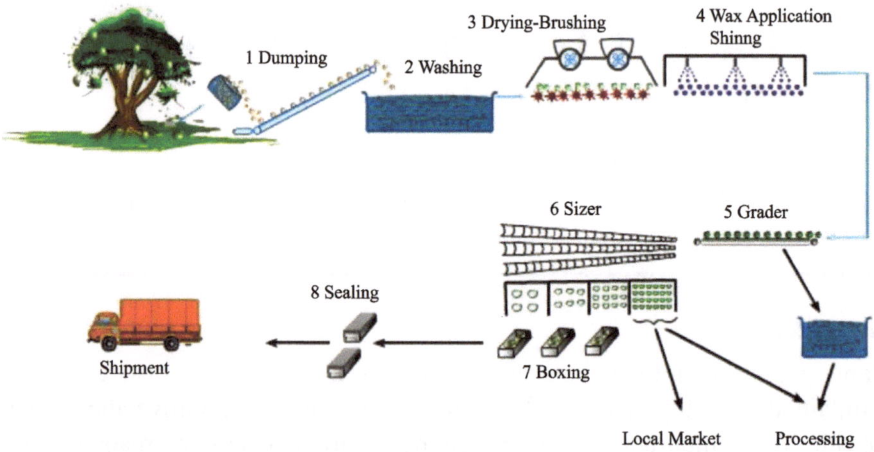

Fig. 2. Schematic representation of post-harvest operation for fresh produce at pack house

Basic conditions for post-harvest treatments are

- Only fresh produce can be preserved
- Produce should be free from defects

A basic principle in shelf life enhancement process is to control or minimize the respiration rate and spoilage.

Post-harvest treatments in handling and storage of horticultural commodities

1. Pre-cooling (Low temperature)
2. Cleaning, washing and trimming
3. Sorting, grading and sizing
4. High temperature - Curing / Drying / Hot water treatments / Vapour heat treatment / Degreening
5. Chemical treatment - Disinfestations/ Sprout suppressants/Mineral application/ethylene inhibitors (1-MCP)
6. Fruit coating (waxing)
7. Astringency removal
8. Irradiation
9. Regulation of ripening -Control/ethylene scavenging/ Degreening

10. Pulsing and tinting
11. Minimal / Light processing
12. Cold storage
13. Packing

Pre-cooling

Pre-cooling refers to removal of field heat (quick cooling) after harvest; if not, its deterioration is faster at higher temperature of 1 hour at 32°C = 1day at 10°C or 1 week 0°C. The entire products must be pre-cooled as early as possible to the recommended storage temperature and relative humidity. Pre-cooling is done just above chilling and freezing temperature.

Advantages of Pre-cooling

- It removes the field heat
- Reduces the rate of respiration, ripening (production of ethylene), the loss of moisture, bruise damage during transits, the growth of spoilage organisms
- Eases the load on the cooling system (refrigeration) of transport or storage chamber
- Above factor helps in extends the product shelf life

Pre-cooling depends on the following factors

- Air temperature during harvesting (during summer pre-cooling time is more)
- Time between harvest and precooling
- Nature of the crop (High perishable crop require immediate pre-cooling)
- Difference in temperature between the crop and cooling medium
- Nature/Velocity of the cooling medium
- Rate of transfer of heat from the crop to the cooling medium.
- Type of package material used - Use of water proof ventilated boxes for good air circulation in the room is helpful. Plastic boxes/ fiber board cartons which have been treated with wax will render them water proof.

Mechanism of pre-cooling

Conduction and convection are the two main heat-transfer mechanisms used for cooling of produce. With conduction, the heat is transferred within a produce to its coldest surface. This is direct movement of heat from one object to another by direct methods (from fresh produce to water or warmer to cooler).With convection, the heat is transferred away from the surface of the produce via a cooling medium such as moving water or air.

Potatoes/ apples/cauliflower/orange and other fruits (bigger mass and lesser surface area) and vegetables require more time to pre-cool than produce which is having smaller mass and large surface area like lettuce/green onion/ carrot tops/peas/corn/brussel sprouts. This is because of the heat from the inside of the crop has to move to the surface before it is transferred in bigger produce.

The rate of cooling depends on individual volume and the exposed surface of product. The difference in temperature between product and the refrigerating medium also needs to be taken into account.

For example: large exposed surfaces, leafy vegetables cool almost 5 times faster than large fruit such as melons (more volume, less surface).

Heat with in the crop comes in two ways

I. Through the convection from the surrounding air mainly from the sun in the form of radiation Eg.: Crop harvested in the early morning will be cooler, since the sun has not been able warm the surrounding air or crop and lower metabolic heat.

II. From the metabolic heat from chemical reactions within the crop (respiration). When the heat is removed by way of evaporative cooling then the fresh produce must not be sealed in moisture proof film like polyethylene bags.

Pre-sorting - Pre-sorting of produce is usually done to eliminate injured, decayed and other unwanted produce before cooling and handling. Pre-sorting will save energy in that culls will not be handled.

Types of pre-cooling methods

A. Cold air - i. Room cooling ii. Forced air cooling (pressure cooling)
B. Cold water / Hydro cooling
C. Top icing - direct contact with icc
D. Evaporation of water from produce
    i. Evaporative cooling
    ii. Vacuum cooling
E. Hydrovac cooling - combination of hydro and vacuum cooling

Commodity -wise cooling methods

Cooling methods	Commodities
Room cooling	All fruits and vegetables
Forced air cooling	Fruits and fruits type vegetables, tubers and cauliflowers
Hydro cooling	Stems, green leafy vegetables, fruits and fruit type vegetables
Package icing	Roots, stems, cauliflowers, green onion, Brussels sprouts
Vacuum cooling	Stems, Leafy and flowers type vegetables
Transits cooling - Mechanical	All fruits and vegetables
- Top icing & channel icing	Roots, stems green leafy vegetables and cantaloupes

A. Room cooling

- In room cooling, heat is transferred slowly from the mass of the produce (by convection) to the cold air being circulated around the stacked containers.
- This is most common and widely used method. Here cold air is passed from the fan and cool by convection process.
- Its commonest use is for products with relatively long storage life and marketed soon after harvest.

  Advantages of this room air-cooling are that produce can be cooled and stored in the same room without the need of transfer and hence it is economical.
- Under this system, cold air from evaporator enters the room, moves horizontally and then passed through the produce containers and return to the evaporator.

  Disadvantage - It takes more time to cool the products- the removal of heat slowly makes this system unsuitable for highly perishable commodities. This is because the product needs at least 24 hours to reach the required storage temperature.
- Almost all crops are suitable for this type of cooling but it is mainly used in citrus fruits, potato, onions, garlic, citrus *etc.*

B. Hydro cooling

Principle- 'The transmission of heat from a solid to a liquid is faster than the transmission of heat from a solid to a gas'. *i.e.* water is better heat conductor than air.

In this method cold water is used for quick cooling of a wide range of fruit and vegetables. Hydro cooling avoids water loss and may even add water to the fruit. Under this method, water is usually cooled by mechanical refrigeration, but ice may be used to make process faster. Chlorine (150-200 ppm)/Iodine/ Nutrients/Growth regulators/ Fungicides can be added in water to sanitize/ improve nutrient status and prevent post-harvest diseases of the produce. For quick cooling of the produce, cold water must constantly be passed over the crop. This can be done by submersing the crop in cold water which is constantly being circulated through a heat exchanges.

Cooling time

- 2 min for asparagus (long & narrow) & Leafy vegetables (more surface to volume ratio)
- 10 minutes for small produce like capsicum (large and globular)/ cherries/tomato
- Up to 1 hour for large products such as melons.

Advantage

Hydro-cooling has the advantage over the pre-cooling method where it helps in cleaning the produce, provides fast, uniform cooling for commodities. It is faster than forced air cooling. Hydro-cooling can be achieved by immersion or through means of a chilled water shower. Not all crops can be hydro-cooled, because they need to be able to tolerate wetting, chlorine, and water infiltration.

Disadvantage

Tank water can be contaminated with microorganisms which can result in increased levels of spoilage during subsequent storage or marketing so chlorine should be added to avoid the problems.

Two types of hydro coolers are generally used

i. **Shower/batch type:** The water showers over the commodity, which may be in bins or boxes, or loosen a conveyer belt. A common design is to transport the crop on a perforated conveyer belt (the speed of the conveyer can be adjusted to the time required to cool the crop) and cold water is pumped from the tank and allowed to fall on the produce in sprinkled type and then falls through

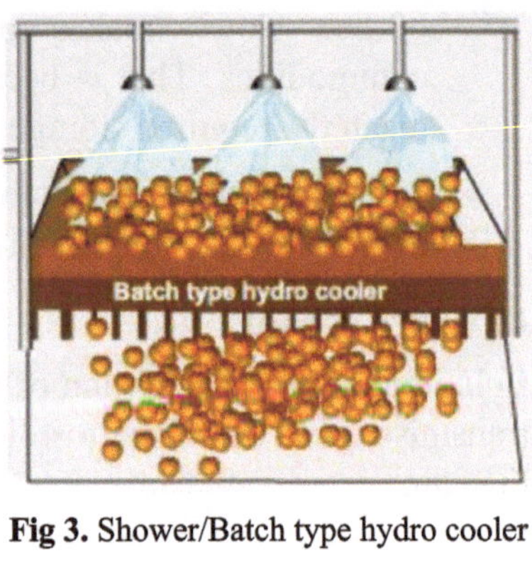

Fig 3. Shower/Batch type hydro cooler

to the tank below then filtered, recycled and re cooled (Fig.3). Efficient cooling depends upon adequate water flow over the product surface. For product in bins or boxes, water flows of 75-100 lt. /min./ft.(400-600 l/ min/m) of surface area are generally used.

ii. **Immersion type:** It is simplest type of a hydro-cooler in which produce is dipped in cold water. Here product are normally in bulk, is in direct contact with the cold water as it moves through a long tank of cold. This method is best suited for products that do no float, because, slow cooling would result if the product simply moved out of the water. Immersion hydro coolers convey product against the direction of water and often have a system for agitating the water. Depth of the water tank should be >30 cm and water tends to penetrate inside fruits, particularly those that are hollow such as peppers. Water temperature also contributes to infiltration. It is recommended that fruit temperature is at least 5°C lower than liquid. Eg.: Radish, Asparagus, Artichoke, Green onion, capsicum and leafy vegetables.

Crops normally hydro cooled

Artichoke	Carrot	Kiwifruit	Potato(early)
Asparagus	Cassava	Kohlrabi	Pomegranate
Beet	Celery	Leek	Radish
Belgian endive	Chinese cabbage	Lima bean	Rhubarb
Broccoli	Cucumber	Orange	Snap beans
Brussels sprouts	Eggplant	Parsley	Spinach
Cantaloupe	Green onions	Parsnip	Summer squash
Cauliflower	J. artichoke	Peas	Sweet corn

C. Forced air cooling or pressure cooling

- In this system 'cold air is passed by force from one side to other side using big fan' (Fig. 4).
- Cold air movement is through the containers rather than around the containers.
- Cooling is 4 to 10 times more rapid than room cooling and its rate depends on airflow and the individual volume of produce
- Air is blown at a high velocity leading to desiccation of the crop. To minimize this effect, air is blown through cold water sprays.
- It is slow compared to hydro cooling but is a good alternative for crops requiring rapid heat removal which cannot tolerate wetting or chlorine of cooling water.

- Adequate airflow is necessary. This is because fruits in the center of packages tend to lose heat at a slower rate, compared to those on the exterior.
- This system is also called as high humidifier. High RH of 90-95% is to be maintained in the precooler to avoid dehydration during cooling.
- This system can be applied to all crops particularly berries, ripe tomatoes bell peppers and many other fruits, cabbage, green peas, cucumber, brinjal, muskmelon, watermelon and mushroom.
    - Baby corn : 5-6 hr cooling at 2-4°C,
    - Leafy vegetables : 1-2 hr cooling at 6-8°C (too low temperature causes chilling injury)

Fig. 4. Forced air cooling (High humidifier)

Difference between Forced air cooling and Vacuum cooling

Forced air cooling	**Vacuum cooling**
The air passes over the surface of the crop, cooling the outside while the inside is cooled by heat transfer from inside to the outside for the crop.	In cooling chamber, pressure (reduced) is exactly the same around the produce and in the centre of the produce. This means the cooling is very even and quick throughout the crop.

Crops usually pre-cooled by forced air

Annona	Citrus	Litchi	Plantain
Avocado	Coconut	Mango	Pomegranate
Banana	Cucumber	Melons	Pumpkin
Barbados cherry	Eggplant	Mushrooms	Rhubarb
Berries	Fig	Okra	Sapota
Breadfruit	Ginger	Orange	Snap beans
Brussels sprouts	Grape	Papaya	Strawberry
Cactus leaves	Grapefruit	Passion fruit	Summer squash
Capsicum	Guava	Persimmon	Tomato
Carambola	Kiwifruit	Peas	Tree tomato
Cassava	Lima bean	Pineapple	Yam

D. Top icing

- This is one of the oldest ways to reduce field temperature.
- It is commonly applied to boxes of produce by placing a layer of crushed ice directly on top of the crop.
- It can also be applied as an 'ice slurry' made from 60% finely crushed ice, 40% water and 0.1% sodium chloride to lower the melting point of the ice.
- Ice slurry give greater contact between produce and ice compared only top icing, and therefore result in quicker cooling.
- The main use for top icing is for road transport and it can be applied shortly after harvest.
- Top-ice on loads should be applied in rows rather than a solid mass. It is important not to block air circulation inside the transport vehicle. Ratios of water to ice may vary from 1:1 to 1:4.
- Direct contact between the produce and the ice provides fast, initial conduction cooling. However, as the ice melts, an air space is created between the ice and the produce and the conduction cooling stops. Subsequent cooling is by radiation and convection, both of which are slower processes than conduction.

Package ice can be used only with

- During transport to maintain a high relative humidity for certain products
- Water tolerant, non-chilling sensitive products
- With water tolerant packages such as waxed fiberboard, plastic or wood.

It also increases costs because of the heavier weight for transportation and the need for oversized packages. In addition to this, as water melts, storage areas, containers, and shelves become wet.

Crops suitable for ice cooling

Belgian endive	Chinese cabbage	Leek	Sweet corn
Broccoli	Carrot	Parsley	
Brussels sprouts	Green onions	Pea	
Cantaloupe	Kohlrabi	Spinach	

E. Vacuum cooling

- Vacuum cooling takes place by water evaporation from the product at very low air pressure (At a normal pressure of 760 mmHg, water evaporates at 100°C, but it does at 1°C if pressure is reduced to 5 mm Hg.).

- Most rapid and uniform methods of cooling. Products that easily/rapidly release water may cool down rapidly.
  - Eg.: Most suitable - Leafy vegetables, cabbage.
  - Not suitable - Tomato with low ratios between mass and surface area and effective water barrier like wax on surface is not suitable.
- Produce is placed in a strong, airtight, steel chamber. Moisture loss is achieved by pumping air out of the chamber containing the product and reducing the pressure of the atmosphere around the product. It causing the water in the produce to vapourize. Cooling occurs because the heat energy for vapourization comes from the produce.
- Vacuum cooling causes about 1% product weight loss for each 5° of cooling.
- This method is also used to cool the products like beans, carrots, capsicum, celery, corn, lettuce, mushrooms, spinach, sweet *etc.*
- High cost and sophistication operation needed.

Preferable Crops for vacuum cooling

Belgian endive	Celery	Mushrooms	Swiss chard
Brussels sprouts	Escarole	Snap beans	Watercress
Carrot	Leek	Peas	
Cauliflower	Lettuce	Spinach	
Chinese cabbage	Lima bean	Sweet corn	

Tips to increase pre cooling- efficiency

- Pre cooling should be done as soon as possible after harvest.
- Harvesting should be done in early morning hours to minimize field heat and the refrigeration load on pre cooling equipment.
- Harvested produce should be protected from the sun with a covering until they are placed in the pre cooling facility.

Precautions

- Since most tropical produce are sensitive to chilling injury, care must be taken not to pre-cool or store the produce below the recommended temperature.
- All produce are sensitive to decay. Pre-cooling equipment and water should be sanitized continuously with a hypochlorite solution to eliminate decay producing organisms.
- Care also must be taken not to allow produce to warm up after pre-cooling. Condensation on pre cooled produce surfaces at higher air temperatures also spreads decay.

Cleaning, Washing, Dressing and Water Spray

Simple postharvest operation such as cleaning, washing and trimming makes produce very fresh after harvest and make convenient for the produce to sales in the market. Preparation for the fresh market starts with dumping onto packing house feeding lines. Dumping may be dry or in water (Fig. 5&6). In both cases it is important to have drop decelerators to minimize injury as well as control the flow of product. Water dipping of produces causes less bruising and can be used to move free-floating fruits. However, not all products tolerate wetting. A product with a specific density lower than water will float, but for the produce which sinks, salts (NaCl) are diluted in the water to improve floatation.

Fig. 5. Dry dumping of lemons Cleaning

Fig. 6. Water dumping of apple

Most of the fruits and underground vegetables like beetroot, radish, carrot, ginger *etc.* when harvested are with soil/mud/latex/pesticides/dried/pest infested/diseases and look dirty. Cleaning and washing makes them marketable. After harvest they should be gently rubbed with wet cloth/dry air and then washed properly to remove all soil and secondary roots. Unclean produce may contain bacteria and fungus which can damage the produce during transit and storage.

- Very small produce is mechanically removed by mesh screens, pre-sizing belts or chains.
- Bruised, rotted, mis-shaped produce, wilted or yellow leaves are usually removed by hand.
- Garlic and onions are topped to remove the dry foliage attached to the bulbs (Fig. 7).
- In many crops soil and loose parts are removed by brushing (Fig.8,9&10).
- For produce such as kiwifruits and avocadoes, dry brushing may be sufficient to clean the produce. Soft brush bristles are preferred in washing, the stiffness must be just enough to remove the dirt without injuring the fruits. The brush speed should not exceed 200 rpm.

Fig. 7. Topping onions before grading

Fig. 8. Brushing and hand removal of damaged fruits before grading

Fig. 9. Spiral wound cylinder brushes

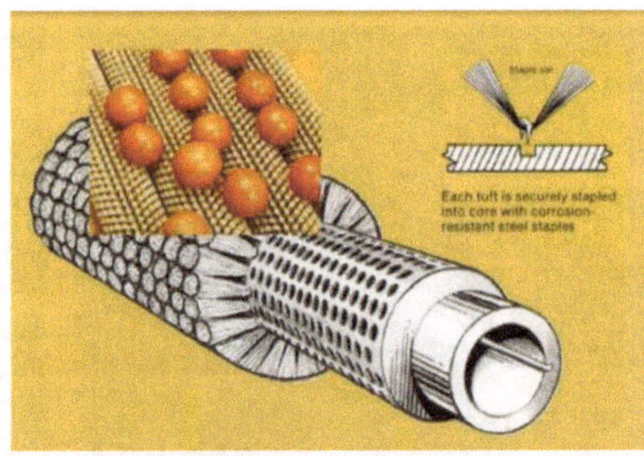

Fig. 10. Staple set cylinder brushes

Washing

- Washing of fruits and vegetables is done to remove adhering dirt, stains insects, molds and sometimes spray residues.
- Washing not only help in cleaning and making the vegetables/fruits fresh and also improves appearance, it also helps in extending the shelf life of the produce.
- Washing is done manually under tap water or in a wash tank using soft muslin cloth.
- Produce should be thoroughly washed with clean water (preferably with 100-150 ppm hypochlorite/chlorine) or soap or calcium hydroxide. Most efficient detergent used is sodium meta bisulphate.
- After washing they are then wiped with dry muslin cloth or air-dried to remove excess surface moisture. Under automated systems, the produce passes under a spray washer on a moving conveyor rollers.
- Thumb rule is to use 1 to 2 ml of chlorine bleach per liter of water gives 100-150 ppm of Cl. pH of the water must be around 6.5 to 7.5.

- Sanitation is essential, both to control the spread of disease from one item to another, and to limit spore buildup in wash water or in the packing house air. Fungicide may be used as post-harvest dip to control diseases and disorder.
- Excess water should be removed from the produce to avoid rotting.
- In crops where water dipping is possible, differential floatation could be used to separate rejects.
- Root and tuber vegetables are often washed to remove adhering soil.

The choice of brushing and/or washing will depends on the type of commodity and contamination

1. Wash before cooling and packing - Carrot (soil), cucumbers, leafy greens, tomatoes
2. Wash to remove latex, reduce staining - mangoes (sap), bananas (debris/ sap)
3. Wash after storage - sweet potatoes, potatoes, carrots
4. Dry brush after curing or storage - onions, garlic, kiwifruit
5. Brushing and wiping - Melons
6. Do not wash - strawberry, beans, melons, cabbage, okra, peas, pepper summer squash *etc.*

Dry cleaning

In some cases cleaning is done by dry brushing instead of washing.

Eg. Removal of white cottony mealy bugs attached in between the surface holes of custard apple fruits.

Some fruits and vegetables are just wiped with clean dry cloth.

Fruits and vegetables which are not suitable for washing are: onion, garlic okra, grapes, strawberry, mushrooms, *etc.*

Dressing

Removal, trimming and cutting of all undesirable leaves/ stem/ stalks/ roots/ other non edible or unmarketable parts is called dressing. Dressing makes vegetables attractive and marketable.

Trimming is done especially in vegetables and flowers to remove unwanted, discoloured, rotting and insect damaged parts (e.g., cabbage, cauliflower, spinach, lettuce, rose, chrysanthemum, gladiolus, tuberose *etc.)* or parts that may favour deterioration or damage during later handling. In case of grapes, trimming of bunches is done to remove the undersize, immature, dried, split and damaged berries. Trimming and removal of decaying parts are preferably done prior to washing. Trimming enhances visual quality, minimizes water loss and other deteriorative processes. Trimming reduces the likelihood of diseases or their spread, facilitates packaging and handling, and reduces damage for other produce.

Water spray

Produce starts loosing water as soon it is detached from the plant. Water spray helps in compensating that water loss and maintaining the quality for longer period. Produce can also be covered with gunny sack soaked in cold water, if it has to store for longer period before sale.

Example: Green leafy vegetables

Sorting, Sizing and Grading

Sorting

Sorting is done by hand to remove the fruits and vegetables which are unsuitable to market or storage due to damage by mechanical injuries, insects, diseases, immature, over-mature, misshapen *etc.* This is usually carried out manually and done before washing. By removing damaged produce from the healthy ones, it reduces losses by preventing secondary contamination. Sorting is done either at farm level or in the pack-houses. In sorting, only sensory quality parameters are taken into consideration.

Sizing

Before or after sorting, sizing is done either by hand or machine. Machine sizers work on two basic principles; weight and diameter. Sizing on the basis of fruit shape and size are most effective for spherical (oranges, tomato, certain apple cultivars) and elongated (Delicious apples and European pears are of non-uniform shape) commodities, respectively.

Mechanisms/Types of sizing

A. Diverging belts/rope grader - the different speed of belts makes produce rotate besides moving forward to a point where produce diameter equals belt/rope separation. Eg. cucumbers, gherkins, pineapples and large root vegetables (Fig. 11).

B. Sizing rollers - with increased spaces between rollers (Fig. 12) Eg. Citrus

C. Sizing by weight - sorting by weight is carried out in many crops with weight sensitive trays. These automatically move fruit into another belt aggregating all units of the same mass. Individual trays deposit fruit on the corresponding conveyor belt (Fig.13). Eg. Citrus, apples and pear and irregular fruits.

D. Hand held template-Sizing can be performed manually using rings of known diameter (Fig. 14).

E. Mesh screens - eg, potato, onion, anola etc.

Fig. 11. Sizing onion bulbs by diverging belts/rope

Fig. 12. Sizing by rollers of increasing distance between them

Fig. 13. Sizing by weight

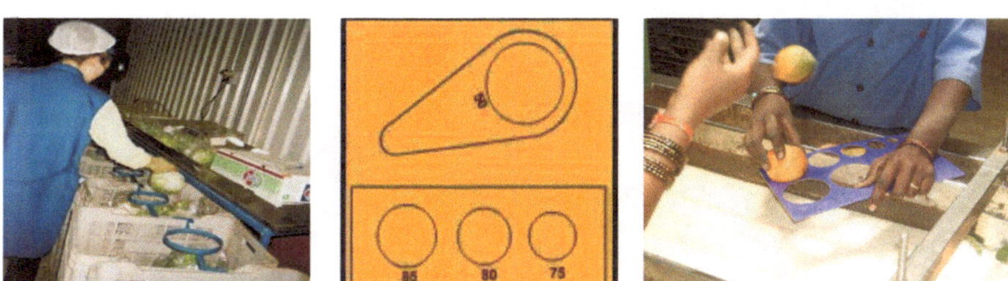

Fig. 14. Sizing with rings of known diameters

Grading

The produce is separated into two or more grades on the basis of the surface colour, shape, size, weight, soundness, firmness, cleanliness, maturity & free from foreign matter /diseases insect damage /mechanical injury.

For eg.: Apple - I. Extra Fancy; II. Fancy; III. Standard; IV. Cull (for processing).

Grading may be done manually or mechanically. It consists of sorting product in grades or categories based on weight/size.

Systems of grading : Static and Dynamic

A. **Static systems** - are common in tender and/or high value crops. Here the product is placed on an inspection table where sorters remove units which do not meet the requirements for the grade or quality category (Fig.12).

Fig. 12. Static quality grading system

B. **The dynamic system** - here product moves along a belt in front of the sorters who remove units with defects (Fig. 13). Main flow is the highest quality grade. Often second and third grade quality units are removed and placed onto other belts. It is much more efficient in terms of volume sorted per unit of time. However, personnel should be well trained. This is because every unit remains only a few seconds in the worker's area of vision. Eg. Onion grading There are two types of common mistakes: removing good quality units from the main flow and more frequently, not removing produce of doubtful quality.

Fig. 13. Dynamic quality grading system

Fig. 14. Grading of Gherkin through divergent rope after deflowering

New Innovation in grading systems

I. Computerized weight grader - operate on the basis of tipping buckets that drops to release the pre weighed item at a particular position. - Apples, citrus

II. Video image capture and analysis-used for size, colour &external defect grading-coffee bean, apple
III. NIR Spectrometers - to assess the TSS non - destructively in apple and stone fruits
IV. X-ray imaging and Computer aided tomography
V. MRI - Magnetic Resonances Imaging
VI. Spectroscopy
VII. Acoustic methods
VIII. Volatile emission analysis

Post Harvets Treatment- Disinfestation

Succulent nature of fruits and vegetables make them easily invaded by these organisms. The common pathogens causing rots in fruits and vegetables are fungi such as *Alternaria, Botrytis, Diplodia, Phomopsis, Rhizopus, Pencillium* and *Fusarium* and among bacteria, *Erwinia* and *Pseudomonas* cause extensive damage.

Losses from post-harvest disease in fresh produce can be both quantitative and qualitative. Loss in quantity occurs where deep penetration of decay makes the infected produce unusable. Loss in quality occurs when the disease affects only the surface of produce causing skin blemishes that can lower the commercial value of a crop.

Disinfestation

Post-harvest diseases of fruits, vegetables and flowers are caused by fungi and bacteria but viruses are rare. These exist either as parasite (on living matter) or saprophytes (dead produce). Most fungi require acidic pH (2.5 - 6) condition in which they grow and develop, while bacteria thrive best at neutral and few can grow at levels below pH 4.5. Bacteria therefore don't usually infect fruits, normally but only vegetables and flowers. Chlorine and sulfur dioxide are most widely used chemicals. Chlorine is probably the most widely used sanitizer. It is used in concentrations from 50 to 200 ppm in water to reduce the number of microorganisms present on the surface of the fruit. However, it does not stop the growth of a pathogen already established.

Mode of infection of micro-organism

Fungal and bacterial infection can occur through mechanical injuries and cut surfaces of the crop, growth cracks and pest damage. They also infect through natural opening on the surface of the crop such as stomata, lenticels, cuticles and hydrathodes. Most fungi are able to penetrate the intact healthy skin of the fruits and vegetables. Many pathogens remain dormant on the surface of the produce for many weeks before visible symptoms of the infection occur.

Damage by micro-organism

It mainly causes physical loss of the edible matter, which may be partial or total. Also affects marketability, particularly where mold growth is obvious on the produce surface. In some cases the superficial infection also make the produce either entirely unmarketable or at least reduce its economic value.

Example: Fungi *Aspergillus flavus* and *A. parasitica* which produce aflatoxin like mycotoxin on ground nut kernels, coconut, dry beans and some leafy foods. The apples juice is also affected by mycotoxin "patulin" produced by *Penicillium patulum* and *P. expansum, P.urticae, Aspergillus clavatus* when stored for too long before being processed. This mycotoxin is carcinogenic and has maximum permitted levels of 50 ppb in fruit juices.

Disinfestations process can be carried out by Physical methods - low temperature, vapour heat and irradiation.

Disinfestation methods

A. **Field management:** Growing fruits and vegetable adopting scientific standard and recommended practices can reduce the field inoculum of disease causing pathogens. Adopting standard cultural practices in terms of sanitation, proper nutrition, irrigation and appropriate harvesting time and methods, etc. are known to reduce post-harvest diseases. Use of wind breaks can reduce spread of field infection where wind is the carrier of pathogen. Other practices like cultivation of crops in regions free from diseases, cultivation of disease resistant cultivars, care in harvesting and handling to avoid wounding fruit, fruit bagging for reducing surface wetness and deposition of inoculum will all help in reducing post-harvest diseases.

B. **Pre-harvest s praying:** Field sprays with fungicides are known to prevent spore germination and the formation of deep seated infections in the lenticels or in the floral remains of the fruits.

C. **Post-harvest chemical treatments:** Post harvest treatments with fungicides like Thiobendazole and Benomyl have rendered good control of stem end rot in many citrus fruits, anthracnose of banana and mangoes despite the fact that infection occurred long before the treatment was applied.

Safer and less toxic chemicals grouped under the category of GRAS (generally regarded as safe) can be used for the control of post-harvest diseases of fruit and vegetables. These compounds mostly include week organic acids, inorganic salts and neutralized compounds. It has been reported that extracts of *Eucalyptus globula, Punica granatum, Lawsonia inumis, Datura stramonium* and *Ocimum*sp. extracts are effective against various fruits rots. Some vegetable and other oils are also effective against fruit rots. Mustard, castor and paraffin oils have been found effective against *Rhizopus* rot of mango.

Disinfection of all handling equipment in pack-houses with 1-3% formaldehyde solution, hypochlorite or SOPP (Sodium ortho phenylphenate) will help in prevention of secondary infection. Washing with water alone reduces many diseases of fruits and vegetables.

Methods of Chemical Application

A. **Dipping** - for effective control of diseases chemical may be used with hot water at 55^{0}C for about 10 min. The crop may be passed below shower of the diluted chemicals. This is called 'cascade' application. Use of chemical like citric acid to lower the pH of the solution along with fungicides seems more effective. In pineapple, infection occur commonly through the cut fruit stalk, therefore dipping cut end was found sufficient to control the disease, save pesticides solution and lower residues on the fruits.

   Eg. Citrus, apples, pineapple, root vegetables.

B. **Spraying** - Spraying is more effective than dipping, because fungicide effectiveness is reduced if the crop has been washed and is still wet and many pesticide chemicals are formulated so that they are not in a solution, but rather in a fine suspension. This results in a concentration gradient in the tank between top (less concentration) and bottom (more) of the tank unless suspension frequently agitated.

   Eg. Citrus, apple

C. **Electrostatic Sprays / Thin film of Coating** - breaking up the pesticides solution into fine droplet and then giving them an electric charge to obtain uniformity of application. Principle is that the particle all have the same electrical charges hence, thus repel each other. These charges are attracted toward the crop and form uniform coating on the produce.

   Eg.: Potato and crown rot of banana

D. **Dusting** - with wood ash and lime in case of yam. Fungicides along with talc on potato.

E. **Fumigation / Vapour treatment** - Fumigation is to eliminate insects, either adults, eggs, larvae or pupae and pathogen inoculum. Fumigant such as sulphur dioxide (SO_2) is used for controlling post-harvest disease in grapes. This is achieved by placing the boxes of fruit in a gastight room and introducing the gas from the cylinder to the appropriate concentration. This treatment results in a residue of 5-18 ppm SO_2 in the grapes is sufficient to control decay. Its toxicity to *Botrytis cinerea* was found to be proportional to temperature over the range of 0- 30^{0}C, where the toxicity of SO_2 increased about 1 to 5 times for every 10^{0}C rise in temperature. In general treatment with 0.5- 1% SO_2 for 20 min is found to be effective followed by ventilation. During storage, periodic (every 7-10 days) fumigations are performed in concentrations of 0.25%.

Disadvantages

a) SO_2 can be corrosive, especially to metals, because it combines with atmospheric moisture to form sulphurous acids. Hence, special sodium metabisulphate impregnated pads are available which can be packed into individual boxes of fruits to gives a slow releases of SO2. Eg.: Grape guard used in grapes fruit packing.

b) At higher concentration it has bleaching effects on black grapes.

c) Some people are allergic to SO_2, particularly those who have chronic respiratory problems.

Litchi fruits- SO_2 fumigation at 1.2% for 10 min. is used to prevent discolouration of the skins of fresh litchi fruits caused by fungal infection, followed by 2 min. dip in 1 N HCl stabilizes the red colour and reduces the skin browning.

Snap beans - Exposing the beans to SO_2 at 0.7% for 30 seconds reduced the broken end discoloration due to mechanical injury. Other chemicals

- Acetaldehydes fumigation in Sultana grapes @500 ppm for 24hr. control postharvest diseases.
- Paper pad impregnated with diphenyl fungicides are commonly applied to citrus fruit.
- Tecnazene, 2-aminobutane (potato) and 2-AB (orange) are the chemicals used.

Fumigation with gaseous sterilants is the most effective techniques for disinfesting produce. However, these are becoming increasingly unpopular or banned because of high mammalian toxicity (hydrogen disulphide), flammability (carbon disulphide) and damage to the atmospheric ozone layer (methyl bromide).

Fumigation with methyl bromide has been replaced by temperature (high and low) treatments, controlled atmosphere, other fumigants or irradiation.

F. Absorbent paper - Chemical may be absorbed into a pad made of suitable material like paper. This absorbent pad soaked in fungicides like thiabendazole and dried, is placed over cut surfaces, such as cut crown end to control the crown end rot of banana. Here pad absorbs latex from the cut surfaces, which also helps to keep the pad in the position and prevents staining the banana. Potassium aluminum sulphate may be added to the pads, which helps to coagulate the latex. This method is used when banana is dehanded in the field and packed directly into export cartoon, where no washing, spraying or dipping take place. Insecticides like dichlorovos has limited vapour phase activity, therefore dichlorovos based pest strip have been included in cartoon packed with flowers to effects ongoing disinfestations during export.

G. Cold storage - Many insect pests do not tolerate prolonged exposure to low temperature. Storing the produce at $<1.6^0$C for 16 days has been shown to be effective for disinfesting fruits against Mediterranean and Queensland fruit fly. But chilling susceptible fruits are not suitable for the this method

H. High temperature - Heat treatments like hot water dips or exposure to hot air or vapor is employed for insect control (and for fungi, in some cases). Using high temperature of about 40-55^{0}C for about 15 minutes can be easily disinfected. Generally, high temperatures can cause softening of tissues and promote bacterial diseases.

- Dipping temperature depends on commodity, insect to be controlled and its degree of development.
- Dipping in hot water also contributes to reduced microbial load in plums, peaches, papaya, cantaloupes, sweet potato and tomato but does not always guarantee good insect control.
- Heat treatments is reconsidered as quarantine treatments in fruits such as mango, papaya, citrus, bananas, carambola and vegetables like pepper, eggplant, tomato, cucumber and zucchinis.
- Temperature, exposure and application methods are commodity specific and must be carried out precisely in order to avoid heat injuries, particularly in highly perishable crops. On completion of treatment, it is important to reduce temperature to recommended levels for storage and/ or transport.
- Many tropical crops are exposed to hot and humid air (40-50°C up to 8 hours) or water vapor to reach a pulp temperature which is lethal to insects. Hot air is well tolerated by mango, grapefruit, Navel oranges, carambola, persimmon and papaya. Similarly, vapor treatments have been used for grapefruits, oranges, mango, pepper, eggplant, papaya, pineapple, tomatoes and zucchinis.
- A common mango fruits disease, anthracnose can be successfully controlled by dipping at 50°C for about 20 min.

I. Biological control

- The yeast *Candida guilliermondii* is used against Penicillium spp. incorporated into citrus waxes
- *Bacillus* subtilis is used against mango anthracnose and stem end rot

Table: List of post-harvest diseases of fruits

Friut	Disease	*Casual organism*
Banana	Crown rot	*Acremonium* sp, *Curvularia* sp, *Colletotrichum musae, Fusarium semitectum, Verticillium* sp.
Ber	Fruit rot	*Alternaria* sp, *Phomopsis* sp, *Colletotrichum* sp.
Citrus	Black rot	*Alternaria citri*
	Grey mold	*Botrytis cinerea*
	Green mold	*Penicillium digitatum*
	Stem-end rot	*Diaporthe citri, D. medusae, D. natalensis.*
Guava	Anthracnose	*Colletotrichum gloeosporioides*
Kiwifruit	Stem rot	*Botrytis cinerea*
	Ripe rots	*Botryosphaeria dithodea*
Litchi	Skin injuries	*Aspergillus* sp., *Penicillium* sp., *Rhizopus* sp.
Mango	Anthracnose	*Colletotrichum gloeosporioides*
	Stem-end rot	*Botryodiplodia theobromae*
	Alternaria rot	*Alternaria alternata*
	Black mold	*Aspergillus niger*
Papaya	Chocolate spot	*C. gloeosporioides*
	Dry rot	*Mycosphaerella* sp.
	Wet rot	*Phomopsis* sp.
	Alternaria spot	*Alternaria alternata*
	Fusarium rot	*Fusarium solani*
	Internal yellowing	*Enterobacter clocae*
	Anthracnose	*C. gloeosporioides, C. dematium, C. capsici, C. circinans, C. papayae*
Pear	Blossom-end rot	*Alternaria* sp., *Botrytis* sp., *Penicillium* sp.
Pineapple	Black rot	*Chalara paradoxa*
	Fruitlet core rot	*Penicillium funiculosum, Fusarium moniliforme, Candida guilliermondi.*
Pomegranate	Heart rot Penicillium rot	*Aspergillus niger, Alternaria* sp.
Stone fruits	Brown rot	*Monilinia fructicola*
Onion	Black mold	*Aspergillus* spp.
	Neck rot	*Botrytis allii*

Table: Chemicals used to control spoilage and quality in fruit and vegetables

Item	Chemicals
Apple	Sodium -phenyl phenate
Banana	Thio bendazole, Benomyl
Citrus	Sodium carbonate, Borax, SOPP, Biphenyl, 2,4 – D; NCl_3 fumigation
Mango	Hot water, Benomyl
Grapes	SO_2 fumigation
Papaya	Hot water
Pomegranate	Ethyloleate
Potato	Hypo chlorite
Carrot & cabbage	Thiobendazole , Benomyl
Onion	Benomyl
Sweet potato & tomato	2,6-dichloro-4-nitroaniline

Post-harvest pests

Although relatively few post-harvest losses of fresh produce are caused by attacks of insects or other animals, localized attacks by these pests may be serious.

- Insect damage is usually caused by insect larvae burrowing through produce, e.g. fruit fly, stone weevil, sweet potato weevil, potato tuber moth and infestation usually occurs before harvest.
- Rats, mice and other animal pests again are sometimes a problem when produce is stored on the farm.

Almost all post-harvest pests originate from field infestations. Wounds and punctures caused by insect pests not only adversely affect visual quality but also serve as entry points for pathogens, leading to secondary infection and spoilage.

Table: List of insect pests affecting postharvest quality

Pests Common Common host		
Fruit flies		
Dacus ciliatus	Lesser pumpkin fly	Cucurbits
D. cucurbitae	Melon fly	Cucurbits and tomato
D.dorsalis	Oriental fruit fly	Most fleshy fruits and vegetables
Mites		
Halotydeus destructor	Red legged earth mite	Leafy vegetables
Panonchus ulmi	European red mite	Apple and other deciduous fruits
Phthorimaea operculella	Potato tuber moth	Potato, tomato, brinjal

Mealy bugs		
Planococcus citri	Citrus mealy bug	Citrus, grape
Dysmicoccus bevipes	Pineapple mealy bug	Pineapple
Moths		
Cydia pomonella	Codling moth	Apple,pear,peach,quince,prunus,walnut
Maruca testulalis	Beam pod borer	Legumes
Scale insects		
Aonidiella aurantii	Red scale	Citrus
Lepidosaphes beckii	Purple scale	Citrus
Quadraspidiotus preniciosus	San Jose scale	Deciduous fruits
Weevils		
Cylas formicarius	Sweet potato weevils	Sweet potato
Sternochaetus mangiferae	Mango seed weevils	Mango

References

Chadha, K.L., Singh, S.K., Ram Asrey, Patel, V.B. and Sethi, S. 2005. Maturity Standards in fruits and vegetables.Proc. First Indian Horticulture Congrss-2004., Vol.II pp.794-817. Published by The Horticulture Society of India, New Delhi.

Chand, R. 2002.WTO, Trade liberalization and Indian Agriculture. Mittal Publications, New Delhi.

Chand, R.2005.WTO and its implication on Horticultural Trade in India. Proc. First Indian Horticulture Congrss-2004., Vo.II pp. 1111-24. Published by The Horticulture Society of India, New Delhi.

Dhatt, A.S. 2005. Post-harvest management technologies for horticultural crops-An over view. Proc. First Indian Horticulture Congress-2004., Vol.II pp. 765-81. Published by The Horticulture Society of India, New Delhi.

Ezekiel. R. 2005. Recent Development in On Farm storage of fresh fruits and vegetables. Proc. First Indian Horticulture Congrss-2004., Vo.II pp.832-43. Published by The Horticulture Society of India, New Delhi.

Finger, J.M. and Nogues, J.J.2004. The unbalanced round outcome. The new area in Future. WTO negotiations. The world economy., 25: 321-40.

Gopal Krishna Rao, K.P. 2005. Influence of pre-harvest treatments on post-harvest quality of horticultural produce. Proc.First Indian horticulture Congrss-2004., Vol. II. pp. 782-93. Published by The Horticulture Society of India, New Delhi.

Ladamia, M.S.2000. Response of Nagpur Mandarin to degreening by ethylene trickle method J Food Sci. Tech., 37: 103-10.

Maini, S.B. 2001. Post-harvest management of fruits and vegetables. In Hand Book of Horticulture (Ed.K.L.Chadha) ICAR, New Delhi pp.930-40.

Meena, R.K. and Yadav, J.S. 2001. Horticulture Marketing and Post-Harvest Management. Pointer Publisher Jaipur, Rajasthan.

Pandey, U.B. 2004. Onion and Garlic storage; Status, Constraints and Challenges. In; Souvenir, First Indian Horticulture Congress-2004., pp. 95-104. Published by The Horticulture Society of India, New Delhi.

Pandey, U.B. 2005. Present Status, constraints and opportunities in Export of fresh vegetables from India. Proc. First Indian Horticultural Congress-2004., Wo.VL pp.965-77. Published by horticultural Society of India, New Delhi.

Pantastico, Er.B. 1975. Post-harvest physiology, handling, and utilization of Tropical and Subtropical fruits and vegetables. The AVI publishing Co. Inc;West port Connecticut.

Saha, N.C.2005. Latest Development in packing of Horticulture produce Proc. First Indian Horticulture Congress-2004., Vo.IIpp. 854-59. Published by The Horticulture Society of India, New Delhi.

Shikhamany, S.D. 2004. Strategies to improve Productivity, Quality and Post-harvest Management in Fruit crops. In; Souvenir, First Indian Horticulture Congress-2004.,pp 5-14. Published by The Horticulture Society of India, New Delhi.

Singh, I.S., Pathak, R.K. 1988. Packaging of Jamun, Aonla and Bael fruits. Souvenir, packaging of fruits and vegetables in India. Agri-Horticultural Society Publication Hyderabad, pp.108-11.

Singh, I.S. 1984. Post harvest handling of fruits (in Hindi). Directorate of Publication Haryana Agricultural University Hisar.

Singh, I.S. 2000. Manual on Post-harvest handling and Processing of fruits. N.D. University of Agri.and Tech. Publication., Faizabad.

Singh, I.S. and Medhi, G1985. Biochemical changes during storage of Beauty Seedless grape. Proc. National Workshop on Post Harvest Management of Grapes., pp.242-48.

Singh, I.S. andWeaver,R.J. 1982.Effect of preharvest application of Ethephon(during stage B3) and S02 generation packing device on Ethylene production and quality of Tokay Grapes during cold storage. Haryana J. Hort. Set, 11: 149-53.

Singh, K. 2004. Strategies for increasing productivity, quality and Post-harvest Management of Vegetables. In; Souvenir, First Indian Horticulture Congress-2004., pp.15-22. Published by The Horticulture Society of India, New Delhi.

Walls, M.M. and Corgan, J.N. 1994. Post-harvest losses from delayed harvest and storage of onion. Hort. Sci., 20: 802-4.

Waskar, D.P. and Roy, S.K. 2002. Zero Energy Cool Chamber storage of fruits. A review. Indian Food Packer., pp.144-47.

4

Cold Chain Management in Food Processing Industry

Dev Raj, A.K. Senapati, J.M. Mayani, N.V. Patel and Avnish K. Pandey

ASPEE College of Horticulture, Navsari Agricultural University, Navsari 396450 Gujarat, India

India is bestowed with huge production of fruits, vegetables, cereals, millets, sugarcane along with meat production, milk production, fish production and poultry. Despite such a huge production in India, owing to perishable nature of these produce around 25-30 percent of the produce is being lost as post harvest loss. The one important reason for this may be the production of the crops are season specific and during the peak season the production is more while demand for consumption is less. It may be possible to cut down these losses to great extent, if these produces may be transported from the produces states to those who are not producing due to agro-climatic reasons. So for making transportation of the perishable a successful venture in decreasing post harvest loss, there must be proper cool chain management facilities both at producing site as well as marketing place. The aim of the cool chain management must be to keep the produce in saleable condition during entire storage periods. Some products are stored for short period whereas fresh fruits, vegetables require cold storage. Foods before keeping into the storage room need some pre-storage treatments for better results.

Pre-Treatments Required for Maintaining Cold Chain

After harvest; most important pre-treatments which are required prior to maintenance of cold chain are pre-cooling and quarantine treatments.

A. Pre-Cooling

Pre-cooling means removal of field heat. It is an important post harvest operation which is accomplished by refrigeration or water immersion with a view to remove field heat, conserve weight and retard ripening and senescence of the harvested vegetables. In short, it is removal of the field heat from the food commodity after harvest. Pre-cooling is prompt cooling after harvest or removal of field heat. *Field heat* is the initial heat of the material (biological objects) and in case of fresh fruit and vegetables, it is the heat of these produce after harvesting from fields which can be determined by the thermal energy removed from the produce in cooling. So, it is the temperature of the commodity at harvest. Pre-cooling refers to rapid removal of field heat from the freshly harvested commodity before shipment, storage/ processing and is essential for many perishable horticultural commodities. Rapid cooling after harvest is referred as pre-cooling. It helps to increase the shelf life of most of the perishable fruits and vegetables. The term pre-cooling was coined by Powell et al in 1904. They observed that there was considerable deterioration during transportation of warm peaches, loaded in refrigerated cars or vans. The cooling of the peaches was so slow that they were over-ripe and decayed by the time of arrival in the market. Then they tried pre-cooling before refrigerated transportation and its benefits were so marked that its practices was successful for other fruits as well. Proper pre-cooling reduces spoilage and helps to retard loss of pre-harvest freshness and quality. Cooling is the removal of heat. Pre-cooling is first step in good temperature management. Delay between high temperature harvest and starts of pre-cooling are certain to increase the deterioration. Pre-cooling at required temperature inhibit growth of decay producing micro-organisms, restrict respiratory and enzymatic activity, inhibit water loss and reduce ethylene production by the product. Most storage room desired to hold products under refrigeration have neither the refrigeration capacity nor air movement needed for rapid cooling. Thus pre-cooling is generally separate operation that requires special equipment and or room.

Benefits of Pre-Cooling

1. Pre-cooling help in reduction in rate of aging.
2. Reduce water loss
3. Reduce physiological weight loss
4. Reduce temperature for refrigeration
5. Heavier loading of transport vehicle.

Factors Affecting Pre-Cooling

There are three main factors for selection of pre-cooling methods.

1. Temperature of the produce at harvest
2. Physiology of the produce.
3. Post harvest Shelf Life of food commodity.
4. Rate of pre-cooling
5. Method of pre-cooling

The commodities that has to be cured at temperature above those required for extending storage is not normally pre-cooled e.g. potato, yam, sweet potato, onions etc. Commodities that have a longer PH life generally do not have to be cooled so rapidly but should be still cooled as soon as possible. Commodities that are susceptible to chilling injury should be cooled according to their individual requirement (10-12°C) (Table 1). Fruits and vegetables where pre-cooling is most important are asparagus, beans, carrots, lettuce, mushrooms, green onions, okra, peas, spinach, strawberry, grapes. Fruits are normally pre-cooled with cold air although stone fruits benefits from hydro-cooling.

Table 1: Chilling and non chilling sensitive commodities depending on physiology

Chilling sensitive	Non-chilling sensitive
Avocado, banana, citrus, guava, papaya, passion fruits, pineapple, sapota, cucumber, Brinjal, musk melon, potato, pumpkin, watermelon.	Apple, Apricot, cherry, fig, grapes, kiwi, pear, peach, plum, strawberry, asparagus, beans, beetroot, cabbage, carrot, cauliflower, garlic, onion, peas, radish, lettuce, brussel sprouts, artichoke.

Factors Affecting Rate of Pre-Cooling

The rate of pre-cooling of produce is dependent primarily up on:

1. Rate of heat transfer from produce to cooling medium, which is especially influenced by rate of flow of the cooling medium around / into the container of the produce.
2. Difference in temperature between the produce and the cooling medium.
3. The nature of the cooling medium.
4. Thermal conductivity of the produce.

When hot produce is exposed to cooled air kept at a constant temperature by refrigeration, the rate of cooling (°C per minute) is not constant, but diminishes exponentially as the temperature difference (driving force) between produce and air falls (Newton's Law). The cooling mediums are Ice, water, or air.

Methods of Pre-Cooling

The choice of pre-cooling method must be related to availability of facility, cost, and types of container, market distance and requirements. After pre-cooling, produce should be refrigerated to prevent losses. Pre-cooling can be done by different methods like

1. Hydro-cooling (cold water cooling by floating methods, spraying and immersion methods).
2. Vacuum cooling
3. Air Cooling (Room cooling, forced air cooling)
4. Contact icing / package icing.

1. Hydro-cooling

Hydro-cooling is a rapid and effective method of cooling. Here water acts as the heat transfer medium. Cooling is accomplished by flooding, spraying and immersion in cold water or enough refrigeration must be supply to keep the water temperature at about 1⁰C or 0⁰C. Commonly used for vegetable and store fruits. The two basic system of hydro-cooling are used i) flow through conveyer system and ii) batch system.

Conveyer Hydro-Cooler: In the conveyer hydro-cooler the product is either showered with or submerged in cold water as it progresses through the cooler. The product may be in a single layer, multiple layers, in bulk bins or in packed boxes. Good immersion hydro-cooler provide overhead showering or flooding at a rate of 400-600 litre/m^2/mint.

Batch Type Hydro-Cooler: The batch type hydro-cooler is normally a room in which palletized unit load of product are stacked 1 to 3 pallets height and cooled with cold water sprayed form overhead nozzle. Hydro-cooling is slower to crated products than for loosened non-packed products because the water does not penetrate to the tightly packed product. Hydro-cooling is preferred for stone fruit preferred.

Hydro-cooling may also clean the produce but contamination of the produce with spoilage micro-organisms occur if soil is not removed from the system regularly and water chlorinated. A further advantage is that the commodity loses little weight during hydro-cooling. Produce and containers must be tolerant to wetting. Prevent re-warming of produce, when cooling is over

2. Vacuum cooling

Vacuum cooling is achieved by enclosing leafy vegetables in an air tight chamber and rapidly pumping out air and water vapour. Water is vaporized in a vacuum chamber under low pressure. Thus cooling is accomplished by evaporation of water from the product surface. As the pressure in the chamber is reduced, evaporation continues. If the pressure is reduced to 4.6 mm of Hg and evaporation continue for a sufficient time at that pressure. At ordinary atmosphere pressure (760 mm of Hg), water boil at 100°C i.e. 212°F. However if pressure is reduced to 4.6 mm Hg, water boils at 0°C. Pre-cooling is done in a vacuum cooler by littering boiling water off the product. Moisture loss during vacuum cooling ranges from 1.5-5%. Since, moisture loss occurs about equally from all parts of the produce being cooled, wilting is not apparent, unless water loss exceed about 5%. (If pressure is increased boiling point of water also increased and vice versa). Boiling point is not co-related with temperature. Some organic solvent boil at very low temperature as boiling is conversion of liquid to vapour. The vacuum chambers are being used very rapidly. The small portable unit have a capacity of a few pallets, whereas the larger stationary one can hold up to two cartloads. Vacuum cooling is the standard commercial method of cooling crisp head lettuce. Rate of cooling and final temperature attain by vacuum cooling affected largely by ratio of surface area of the commodity to the mass. The ease with which product give H_2O from its tissue and the rate of which vacuum is drawn in the chamber determine the affectivity of vacuum pre-cooling. The fruit and vegetables differ widely in their adaptability of this pre-cooling method. Leafy vegetables with large leaf area for evaporation can be pre-cooled easily by this method e.g. Lettuce, spinach, parsley etc. Some other as asparagus, broccoli, cauliflower, cabbage, celery etc may also be pre-cooled by this method if they are rewetted / misted to supply some of the water needed for vacuum cooling and consequently reduce the amount of the moisture loss from within the tissues. Products with low surface to mass ratio such as trees fruits, most fruit crops and tomato are better adapted to other cooling methods

Vegetables that have high surface to volume ratio may be rapidly and uniformly cooled by boiling off some of the water at low pressure. Vacuum cooling is as rapid as hydro-cooling. Here the produce is loaded into a sealed container, and the pressure is reduced to about 660 pascals (5 mm of Hg). At this pressure water boils at 1°C and the produce is cooled by evaporation of water from the tissue surface. For every 5°C drop in temperature approximately 1% of the produce weight is boiled off as water. The weight loss may be minimized by spraying the produce with water either before enclosing it in the vacuum chamber or towards the end of the vacuum cooling operation (Hydro-vacuum cooling). The rate of cooling by this technique is largely dependent on the surface to volume ratio of the produce and the ease with which the produce losses water. Leafy vegetable are suitable for vacuum cooling (Table 2). Strawberries are suitable for vacuum pre-cooling. Some product like cantaloupes can be pre-cooled by several ways but not by vacuum.

Table 2: Decrease in temperature due to similar vacuum cooling condition in some vegetable.

Produce (initial temperature 20°C)	**Final Temperature (°C)**
Lettuce, onion	2
Sweet corn	5
Broccoli	6
Asparagus, cabbage, celery, pea	7
Carrot	14
Zucchini	18

3. Air cooling

Air cooling is accomplished by using either room cooling or forced air (Pressure) cooling. Water loss can be a problem in air cooling but can be almost eliminated by providing high humidity air. Relative humidity of 95% or above is used in the air cooling.

i) **Room cooling:** It is most common cooling technique. The produce in boxes, carton, bulk or other packages is exposed to cold air in a normal cool store i.e. in the same refrigerated space used for temporary or long term storage. For adequate cooling air velocity around the package should be about 60-120m/ mint. Container must be properly stacked to permit air flow between them.

Disadvantages

- More space is required
- Rate of cooling is relatively slow and thus may be inadequate for more sensitive produce.
- Disadvantage is more if field bins or container are utilized on pallets.

ii) **Forced air cooling:** The rate of cooling with cold air may be significantly increased if the heat transfer surface is enlarged form that of the package to the total surface of the produce. It is accomplished by producing a difference in air pressure on opposite face of stacks of ventilated containers. This method is commonly used in USA and Australia. The pressure difference between opposite faces ranges from barely measureable to about 250 pascal (25 mm water head) and air flows vary between 0.1 and 2.0 lt/sec//kg. The speed of cooling can be adjusted by varying the rate of air flow. This difference forces air through the stacks and carries produced heat away. Forced air cooling usually takes only 1/4th to 1/10th the time needed for room cooling but still takes twice as long as hydro-cooling or vacuum cooling. Room cooling removes heat from only the surface of package, the size and shape of the package being the limiting factors. Forced air cooling after all is the solution for cooling fresh fruits, vegetables, and melons (even cut flowers) that cannot be hydro-cooled. The refrigeration requirements for forced air cooling are often over estimated because of a lack of understanding of the factors that limit the rate of heat loss from the cooling produce.

4. Package Icing (contact icing)

Package icing is an old method that involves placing finally crushed ice with in shipping container. It can be effective in cooling products that are not harm with contact with ice. It is now mainly employed as a supplement to other forms of pre-cooling. The finally crushed ice or ice slurry (liquid ice containing about 40% water and 60% ice and 0.1% salt) is sprayed on to the top of the load inside the road or rail transit vehicle. This is often referred to as top icing. Addition of salt reduces the freezing point (-0.5^{0}C instead of 0^{0}C). In Australia, package icing is commonly used for cooling broccoli and sweet corn packed in polystyrene containers to maintain freshness during transport and marketing.

5. Evaporative cooling

This is a simple process in which dry air is cooled by blowing it across a wet surface. Although the technique is restricted to regions with low RH but with a good quality water supply, it has the advantage of low energy cost. The extent to which air may be cooled by evaporation of water is limited by the water holding capacity of the air.

B. Quarantine Treatments

Fumigation is done to check the insects. Efforts should be made to keep the pest out in production areas of country which are free from certain crop infesting insects/diseases. The commodities subjected to treatment for disinfestations must be certified as free from the insect and disease. A number of treatments have develop for sterilization of plant material to ensure that crop damaging insect and diseases can not spread from an infested area to a main production area. Various quarantine treatments includes fumigation, vapour heat treatment, cold treatment, pre-packaging and irradiation.

i) **Fumigation treatments/ fumigants:** Mainly used for insects. Earlier, CS2 is use but now banned. HCN gas also banned at present. Two fumigants like methyl bromide (MB) and ethylene dibromide (EDB) was used for disinfestations in USA. Both of these are effective and reasonable safe to use. MB is more toxic to many fruit than EDB. So the later is generally recommended for citrus and stone fruit. MB is used for disinfection of potato, sweet potato and tomato. Amount of MB and EDB varies from 5-3 lbs/ 1000fts3 and treatment time varies from 2-4 hrs. SO_2 (KMS-potassium metabisulphite) is also used commercially. SO_2 is standard practice for controlling decay of grapes. It also retard respiration and preserve the fresh colour of commodities. Biphenyl is impregnated in papers pad or fruit traps and used to control decay in package of citrus fruit. It inhibits sporulation of penicillin on decaying fruit thus prevent spoilage of adjacent fruit by mould spores.

   Chlorine 50-100 ppm is often used in wash water/soap tanks/hydro-cooler to check the micro-organisms. There is one chemical Sodium

Ortho Phynyl Phenate used with detergent is helpful in removing spores from fruit and reducing the population in washing water. Fungicides are also applied with cleaning, waxing such as Thiobendazole (TBZ), 2-amino butane, benomyl are some fungicides. Some other fungicide like diphenyl amine and Ethoxyquin usually give good control of storage scald of apple. Growth regulators (GR) are used to control sprouting during storage e.g.: Onion, potato. Malic hydrazide (MH) and MENA (methyl ester of naphthalene acetic acid) are most commonly used GR to prevent sprouting of onions. Lemon treated with 2,4 diphenyl dichlorphenoxy acetic acid (2,4-D) used before storage to delay alternaria rot and senescence of the calyx. Calcium chloride is used to control bitter pit in apple. It helps to maintain firmness. EDB consumption in USA is prohibited from 1st September 1984.

ii) **Vapour heat treatment (VHT):** In this method the temperature of the fruit is gradually raised by saturated water vapour at 43oC until the centre of fruit reaches the temperature in 8 hrs. Fruit is held at this temperature for 6-8 hr. This method is used as alternative method of fumigation in citrus fruit, mango, papaya, and few vegetable imported to the united state from the infested country.

iii) **Cold treatment:** Another choice for meeting quarantine is refrigeration of produce for a specified time because certain fruit infesting insects will not survive at low temperature. In this method the fruit require continuous exposure of temperature of 2°C or lower for periods upto 22 days exception grapes. In case of tropical and subtropical fruit this treatment is not given. This is given to deciduous fruits.

iv) **Pre-packaging:** To pack the material in consumer size packages and to pack a portion required for consumption.

v) **Irradiation:** Treating produce with rays. delays ripening and commercially used in Maharashtra for treating onion. In minimize insect spoilage, microbial spoilage, control sprouting and rooting in onions, garlic, potato during storage (Table 3). It has potential as disinfestations treatment. It is able to control citrus fruit fly, mango seed weevil, and papaya fruit fly. Alphonso, a popular mango variety for export propose are treated with 25 krad dose of radiation.

Table 3: Irradiation effect on storage life of important fruits and vegetables

Commodity	Dose krad	Storage life (Ambient temperature
Alphonso Mango	25	6-8 days Delayed ripening
Papaya	75-100	6-8 days Delayed ripening
Banana	40	4-6 days Delayed ripening
Avocado	35	24 days delay in repining
Guava	30	3-5 day delay in ripening
Onion, garlic, potato	5-15	Inhibit sprouting

C. Chemical Treatments

Some time before storage food commodities are being treated with chemicals to increase shelf life of perishables.

1. Spray 0.5% calcium chloride (0.5 kg/100 litre water) starting from 1st/ 2nd week of July at two weeks interval for at least three times for the control of bitter pit and other calcium related disorders. Addition of Bavistin @ 0.05% to the spray solution in the last spray remarkably reduces the incidence of blue mould (Penicillium expansum).
2. Apple fruit can also be enriched in calcium by PH dips
   i. Dip the fruits in 4% calcium chloride solution + 0.03% surfactant such as Uphar or Tween 20 or Tween 80 for 1-2 minutes followed by air drying and storage.
   ii. Hydro-cool warm fruit at field temperature in chilled (3 to 4°C) 2% calcium chloride solution containing 0.03% surfactant for 20 min for rapid calcium infiltration without any risk of Ca injury
   iii. Boric acid sprays given @ 0.1 to 0.2% before 45 -60 days of harvest significantly improves the calcium status of apple at harvest.
3. A single spray of silver nitrate @ 75 ppm or DNP @ 30 ppm 6 weeks before harvest increases fruit firmness at harvest and hence storability in apples.
4. PH dip in diphenylamine @ 0.1 to 0.25% for 30 seconds can effectively control superficial scald of apple
5. Enclosing a sachet of $KMNO_4$ crystals (3 to 5 g/l to 2 kg fruit) in polythene bags (100 gauge) with 0.5% venting area results in better quality retention and storability in pre-packaged apples.
6. Alar @ 1000 ppm sprayed 60 days before harvest delays the ripening and improves the storability of apple and pears.

7. Santa Rosa plum fruit given a postharvest dip in 4% $CaCl_2$ for 2 mint results in high colour development and reduction in loss of flesh firmness during 12 days ambient room storage.
8. Pre-packaging mid hill apples or Sants Rosa plum in sealed polythene bags of 150 gauge thickness without ventilation results in better quality retention and longer storability
9. Shelf life of kinnow fruit can be increased by giving a PH dip in 0.1% bavistin for 5 min followed by packing of 5-10 fruits in sealed polythene bags extends shelf life upto 6 weeks under ambient conditions and 10 weeks under evaporative cool chamber.
10. PH treatment with GA markedly retards ripening of tomato and bananas.
11. A commercial preparation of ethylene absorbent called Purafil, proved effective in the complete absorption of ethylene from banana held in sealed polyethylene bags.

Storage Methods for Management of Cool Chain

Preservation of food by removal of heat means the food storage of by use of low temperature. Microorganisms are not killed by removal of heat (low temperature), but their multiplication and activities are inhibited. Low temperature also retards chemical changes as well as enzyme reactions. Lower the temperature, the greater is the retardation of the microbe. Huge quantities of eggs, meat, fruits and vegetables are kept in cold storage, so that they can be made available throughout the years to the consumers. In all the cases the principle involved is the same viz., the temperature inhibition of microbiological and biochemical actions (deteriorative actions) by low temperature. Low temperature can be produced by (i) cold storage (low temperature, high RH) (ii) Controlled Atmospheric (CA) storage (low temperature, high RH, decreased /low O_2, high CO_2) or Chill storage (-1°C to -4°C) (iii) Hypobaric and hyperbaric storage (iv) Zero energy cool chamber (ZECC) (v) Cool night air (vi) Cellar storage (about 15°C), (vii) Refrigeration or chilling (0 to 5°C) (viii) Freezing (-18 to -40°C) (ix) Dehydro-freezing (beyond -40°C) (x) Modified atmospheric (MA) storage. These methods can be categorized into low cost storage system or high cost storage systems.

A. Low Cost Storage Methods / Cost Effective Storage Systems

1. On-tree Storage
2. Home storage
3. In-Ground Storage
4. Ice refrigeration

5. Air Cool Storage
6. Zero Energy Cool Chamber (ZEEC)

B. High Cost Storage Methods

1. Mechanical Refrigerated systems
2. Gas Storage: i) Controlled Atmospheric (CA) storage and ii) Modified atmospheric (MA) storage
4. Pressure storage: i) Hypobaric storage (low pressure) and ii) hyperbaric storage
5. Refrigerated Transport (Vans)

A. Low Cost Storage Technology / Cost Effective Storage Systems

1. On-tree storage system: It is also called Onsite storage. It means delaying harvest until crop is required for consumption. It can be done in certain cases with root crops. It is also called in-situ storage. Application of GA_3 spray @ 10-15 ppm before maturity increased storage life of mango fruits on- tree it shelf due to delayed ripening in mango.

2. Home storage: Green vegetables cannot be stored well for more that 2-3 days during winter while in summer their shelf life further decreased because of higher temperature. Potatoes can be stored by spreading over the floor of a cool and dry room with few cm of fine sand whereas onions can be stored by putting them in thin layers over the floor along with garlic. Store room should be properly ventilated. Onions, sweet potatoes, squash and pumpkin keep better if stored in shelves or stacked crates than when spread over the floor.

3. In-ground storage: These are more sophisticated forms of below grown storages, they may be part of above ground building or underground rooms, often in hillsides, but good drainage and protection from rain are essential. The cellars are cooler than above ground building in summer, warm weather and warmer in winter, reducing the risk of freezing Hard vegetable and long keeping fruits can be stored eg: apple. Method is commonly practiced in China.

i) **Pit storage or clamp storage:** Pit storage or clamp storage is still practiced in some developing countries. Hard vegetables such as potato, turnip, jimikand (eleplant foot), ginger and late season cabbage; are piled into pits dug into a hillside or other well drained spot. The pits are lined with hay or straw and produce is then covered with straw followed by 10-20 cm sod and earthed to protect against freezing and rains. Piped ventilation is provided outside. Modified pit system is developed in Solan (India). Outside ventilations help to reduce respiratory self heating. Clamp storage is used for storage of cassava for upto 2 month.

ii) **Cellar storage (about 15°C):** The temperature in cellars (underground rooms) where surplus food is stored in many villages. The storage is usually not much below that of the outside air and is seldom lower than 15°C. It is not enough to prevent the action of many spoilage organisms or of plant enzymes. Decomposition is, however, slowed down comparatively / considerably under cellar storage. Root crops, potatoes, onions, apples and similar foods can be stored for limited periods during winter months. The performance of cellar is improved by providing controlled ventilation opening for entrance of cold air and exit of warm air by conventional circulation when cooling is required. Cellarage is an underground space especially for storage is known as cellarage.

4. Ice refrigeration: Ice is kept in storage chamber in ice box or ice chest. It is used to preserve/store meat and other perishable foods. Harvested ice should be kept in insulated 'ice houses'. The melting of 1 kg ice absorbs 325 K.J/ of heat.

Disadvantage: i) More ice is needed and problem of melted water disposal.

5. Cool night Air (Air cool storage): These are simply insulated structures above ground or partly underground which are cooled by circulation of colder, outside air. This low cost system of storage can be used advantageously at places where there is a wide different between the day and night temperature. When the temperature of the produce is above the desired level and if the temperature of the outside air is lower, air is circulated through the store by conventional or mechanical means via bottom inlet vents and top outlets fitted with dampers. Air cooled storage is cheap to construct and easy to operate.

Sand heated very fast in day and cools very fast in night and so harvest the fruits in night. In north India when nights are very cool, then blow the cool air from outside to inside and in this way fruit and vegetables are stored. In day time temperature increases, then thermostat blow the hot air from inside to outside and cool air to inside. Potato and sweet potato, both of which need relatively high storage temperature to avoid accumulation of sugar and chilling injury respectively. Apple can also be store in high hill above 2000 metres.

Air storage: Storage in the room at normal air temperature providing proper air flow and ventilation which facilitates the removal of ethylene produced by fresh fruits and vegetables and maintains the CO_2 and O_2 level about 0.03% & 21.0%, respectively.

Dry storage: Storage of food commodity without extremes of temperature change, at about 20°C and below 50% relative humidity.

6. Zero energy cool chamber (ZECC): Zero energy cool chamber (ZECC) is used to enhance the shelf life of fresh fruit and vegetables by maintaining the temperature and RH during summer and winter season naturally with the application of water. It is based on the principle of direct evaporative cooling *(evaporation cause cooling effect)* and hence, does not require any electricity or power to operate and all the materials (Brick, Sand, Bamboo, Khaskha, etc.), required to make the cool chamber are available easily, cheaply and are reusable. ZECC constructed by bricks and bricks are not cemented. Chamber is covered by polythene sheet to prevent sun rays and rain water. Put a drum of water above wall and drip the water to sand by pipe. Water evaporate from wall and energy required for evaporation comes from inside the room and room get cool and in this way temperature may decrease upto 10^{0}C i.e. 35^{0}C – 25^{0}C and in this way can store the fruit/vegetable without energy source also (Table 4). This method is given by Australian and now well adopted in India. Problem is that at the bottom there is lot of water which cause PH disease and treatment with TBZ @ 1000ppm, even can solve the problem. In this way huge loss of monitory value due to power cut in storage in potato every year in India can also be minimized.

In India, it was developed by S.K. Rao and Khuradia in 1981 at Indian Agricultural Research Institute (IARI), New Delhi. In Australia the ZECC was developed in 1960. Its main advantage is that it is low cost technology; as electricity is not required. This chamber can be used effectively for storage of fruits and vegetables. Cooling occur when unsaturated air is blown with water vapour across any wet surface. It maintains uniform high RH (85-90%). Increase in shelf life of fruits and vegetable in ZECC is not only because of lower temperature but also due to uniform high humidity. The storage life can be increased 2-3 times as compared to ambient condition, especially during summers. This method is feasible where the temperature during summer is very high and relative humidity (RH) is very low. So, ZECC can be used in lower hills and valley areas. It reduces temperature by 12-14°C. The raw materials used in the preparation includes 400 bricks, 10 bags river bed sand, bamboo, gunny bags.

Design and construction: The floor of the storage space is made with a single layer of bricks. Over this a double walled rectangular structure is erected with approximately 7.5 cm space between the inner and the outer bricks walls. The outer dimension of chamber should be about 165 cm x 115 cm x 67.5 cm. The cavity between the two walls is filled with riverbed sand. The top of storage space is covered with khaskhas/gunny cloth in a bamboo frame structure. It should be near water source and there should be lot of aeration. The chamber should be kept wet by sprinkling water once in morning and once in even to maintain low temperature and high RH.

Precaution: Saturation of water should be done regularly but excess water should be avoided. It should be clean and free from insect pest and diseases. Disinfect the cool chamber periodically by spraying/sprinkling suitable disinfectants such as chlorine water or bleaching powder or by sulphur fumigation.

Table 4: Storage life of different fruits and vegetables in ZECC

Fruit/Veg.	Months	Cool chamber Temp. (^{0}C)	Storage life (days)
Plum	June	4	10
Mango	July	8	15
Ber	March-April	4	10
Kinnow	Jan-March	8	60
Cauliflower	Feb.-March	7	12
Peas	Feb.-March	5	10
Cucumber	May-June	3	8
Carrot	Feb.-March	5	12
Radish	Feb.-March	5	10
Ladies finger	May-June	1	6
Spinach	Feb-March	3	8
Bitter gourd	May-June	2	6
Green chillies	May-June	3	6

B. High Cost Storage Technology

1. Mechanical refrigerated system: Australian Engineer James Harrison is known as Father of modern refrigeration because he developed first practical refrigeration system in 1851. Refrigerator is an apparatus / room or appliance that is used for keeping food or other items in cool / refrigerated conditions. It is an insulated cool camber in which edibles, medicines etc are kept to preserve their freshness. Refrigeration plant consist of 4 basic component: (i) the compressor in which the refrigerant gas, either ammonia or halogenated hydrocarbons (Freon-12) is compressed (ii) the condenser, either air cooled or water cooled, in which the hot gas is cooled and condensed to a liquid, (iii) the expansion value (iv) the evaporator coils in which the liquid is permitted to boil and so remove heat from its surroundings. The motor compressor compresses a suitable gas (Refrigerant) which liquefies and expands in a closed system and in this process it absorbs the latent heat from the surrounding resulting in cooling of the chamber.

Refrigeration may be defined as the process of achieving and maintaining a temperature below that of the surroundings, the aim being to cool some product to the required temperature. One of the most important applications of refrigeration has been the preservation of perishable food products. Thus, refrigeration is removal of heat. It is commercial system used in India for storage of potatoes.

Further, refrigeration is a process to make or keep cold / freeze / chill a food product for preservation. Or Refrigeration is the art of producing cold which is done by removal of heat from the insulated chamber by means of refrigerants. It is the process of lowering the temperature of substance to that of its surrounding atmosphere by using refrigerant. Refrigerated storage is a storage method at a temperature below those normally available throughout the year. Refrigerated storage is generally done between about 0 to 8^{0}C. The foods must be stored above collapse temperature. Collapse temperature is the temperature at which food material losses its structure and volume during storage and processing. Different refrigeration systems are enlisted ahead.

i) **Refrigerated storage or chilling storage:** Chilling temperature (0-5^{0}C) are obtained and maintained by means of ice or mechanical refrigeration. Fruits, vegetables and their products can be preserved for a few days to many weeks when kept at this temperature. The best storage temperature for many foods is slightly above 0^{0}C but this varies with the product and is fairly specific to it. Besides temperature; the RH and the composition of the air can affect the preservation of the food. Generally, refrigeration operates at 4 to 7°C.

   Ton of refrigeration: The standard unit of refrigeration capacity which is derived on the basis of the removal of the latent heat from 1 ton of water at 32^{0}F in order to produce 1 ton of the ice at the same temperature in 24 hrs. Since the latent heat of Fusion for water is 144 BTU /lb, a ton of refrigeration is 144 times 2000 lb or 288000 BTU.

ii) **Cold Storage:** Storage of food in a cold place for preservation. It is an insulated storage with refrigeration to maintain a stable cold temperature for long term storage of perishable products. A building made for cold storage of the food commodities is called cold store. In cold storage low temperature ranging from -1^{0}C to 15^{0}C and RH is 85-90% is maintained depending in food and therefore, there is no exchange of moisture and thus prevent shrinkage of fruit and vegetables. Different fruit and vegetables have different storage temperatures (Table 5 to 7). The temperature at which a food can be stored depends upon the type of the food as well as economic factors. Cold storage is an isolated storage structure with refrigeration to maintain a stable cold temperature for long term storage of perishable product. Cold storage is the principal method of preservation and act as an adjunct to other methods of preservation. Maximum acceptable level of water loss in fruits and vegetables during storage is presented in Table 8.

Commercial cold storage with proper ventilation and automatic control of temperature are now used through the country of the storage of semi perishable foods such as potatoes and apples. Mesophilic organism can not grow in refrigerated conditions because they are unable to grow below 10°C, but some mesophilic micro-organisms are psychrotropic and can grow on refrigerated foods. Psychrophilic micro-organisms grow in refrigerated foods only. However, the optimums growth temperature of these micro-organisms is 25°-30°C and reduced growth occurs at temperature of 10°C or less and cause spoilage of food. Food will spoil about four times as fast at 10°C and twice as fast at 5°C as at 0°C. Some micro-organisms causing food borne illness are psychrotrophic and but most will not grow or produce toxin below 4.4°C. None of these micro-organisms will grow below 1.7°C. Quit often, other systems for preserving foods are used in conjunction with refrigeration. Since psychrophiles are primarily aerobic, vacuum or CO_2. Packaging with the exclusion of oxygen will delay micro-organisms on refrigerated foods.

- **Cold shock:** If the food products are cooled rapidly from an optimal temperature to 0°C, the micro-organisms present in foods may die which is known as cold shock. There is reduction in temperature without freezing of the substrate. Cells in the early exponential growth phase are most sensitive to cold shock. Cold shock results in loss of one or more membrane permeability barriers. Hence, there is a leakage of amino acids and nucleotides from the cells.
- **Rapid chilling:** It is necessary to remove the field heat form fruit vegetable the animal heat form milk, eggs or meat to prevent spoilage of these foods. Foods are rapidly cooled by forced air cooling, spray chilling, hydro-cooling, and icing or in continuous chillers in slush ice (thawing muddy snow). Fluidized beds are used for cooling vegetable after blanching and prior to freezing. Food should be refrigerated in small containers, so that it can be cooled with in short time. Bulk refrigeration can be a contributing factor in food borne illness. To allow air circulation and to maintain uniform temperature in store room, food should not be stacked tightly.

Table 5: Cold Storage temperature of Fruits and Vegetables

Storage Temp °C	Commodity
	Vegetables
0°C	Globe Artichoke, Asparagus, Beetroot (topped), Broccoli, Brussels sprouts, Cabbage, Carrots (topped), Celery, Cauliflower, endive, garlic, Leafy greens, leek, Lettuce, Mushrooms, Rhubarb, Shallots, Silver beet, Sweet corn, Turnips (topped), Onions, parsnips (topped), peas, spinach, Bitter gourd, coriander, colocasia, Radish
4-7°C	Beans, Capsicum, cucumber, Eggplant, Brinjal, Okra, Potatoes, Rock melon, Squash, Tomatoes (red ripe), Watermelon, Zucchini, ginger, Green pepper
8-13°C	Marrow (hard), Pumpkin, Squash (hard), Sweet potato, Tomato (coloured)
	Fruits
0°C	Grapes, Figs, Litchi, Pomegranate, Apple, Apricot, Cherry (sweet), Peach, Pear, Persimmon, Plum, Quinces, Strawberry
7-9°C	Citrus, Chiku, Custard Apple, Dates, Papaya
9-13°C	Guava, Mango, Papaya, Pineapple

Table 6: Recommended temperature and relative humidity for cold storage of vegetables

Sr. No.	Vegetables	Temp. (°C)	RH (%)	Storage life
1	Asparyagus	0.0	95	3-4 (weeks)
2	Beans, dry	4-10	40-50	6-10 months
3	Beans, green	4-7	95	7-10 days
4	Beets, bunched	0-.	90 (98-100)	10-14 days
5	Beets, topped	0.0-0.7	90-95	8-14 wk. (4-6 m)
6	Bitter gourd	0.6-1.7	85-90	4 wk
7	Brinjal	10.0-11.1	92	2-3 wk
8	Broccoli	0.0	93-100	10-14 days
9	Cabbage, early late	0.0-1.7	92-95	4-6 wk
10	Carrot, mature topped	0.0-1.7	98-100	12 wk
11	Cauliflower (snow ball)	0.0-0.7	85-95	7 wk
12	Celery	0.0-0.6	92-95	8 wk
13	Colocasia	11.1-12.8	85-90	21 wk
14	Coriander leaves	0.0-0.7	90	5 wk
15	Cucumber	10.0-11.7	92	2 wk
16	Egg plants	8-12	90-95	1 wk
17	Garlic	0.0	65-70	28-36 wk (6-7 m)
18	Ginger	7.2-10.0	65-75	16-24 wk (6m)
19	Hyacinth bean, pod	0.6-1.7	90	3 wk
20	Lettuce, leaves	0.0	90-95	1.5-2

21	Lima bean, pod	4.4-7.2	90-95	1.5
22	Musk melon (Cantaloupe) (Honeydew)	1.7-3.3 7.2	85-90	4.5
23	Mushrooms	0	85-95	3-4 day
24	Okra	7-10	90-95	2 wk
25	Onion Green Dry (bulb)	0	95-100	3-4
26	Peas, green	0.0	63-70	20-24
27	Peppers Greek, ripe	0.0	88-92	34
28	Potato	7-2	85-90, 90-95	5-10
29	Chilli dry	5.6-7.2	60-70	6
30	Pumpkin	3.0-4.4	70-75	24-36
31	Radish topped winter	0-10	88-92	3-5
32	Squash (Winter)	1.7-11.6	70-75	24-36
33	Spinach	12.8-15.6	95-100	10-14
34	Sweet corn	0.6-1.7	90-95	13-20
35	Sweet potato	10.0-12.8	80-90	4-7
36	Tapioca, root	0.0-1.7	85	23
37	Tomato Unripe (Mature green) Firm ripe	13-21 (8.9-10.0)	90-95	3-4
38	Turnip	8-10 (7.2)	90-95	8-16
39	Water melon	7.2-15.60	80-90	2
40	Yam	26.7	66-70	3-5

Table 7: Recommended temperature and relative humidity for cold storage of fruits

Sr. No.	Fruits	Temp. (^{0}C)	RH (%)	Storage life
	Tropical and subtropical fruits			
1	Banana (Green for ripening)	12-13	85-95	1-2 (weeks)
2	Banana (Ripe)	12-13	85-95	2-3 (weeks)
3	Citrus			
	i) Mandarins			
	Coorg Orange	6-8	85-90	10-12
	Nagpur Orange	5-6	85-90	10-14
	Punjab santra	2-5	85-90	4-6
	ii) Sweet orange			
	Malta, blood red	2-5	85-90	16
	Malta, common	5-6	85-90	16
	Valencia late	5-6	85-90	20
	Mosambi	7-8	85-90	16
	Sathgudi	7-8	85-90	16

	iii) Pummelo, lime	8-10	85-90	3-6
	iv) Lemon	8-9	85-90	8-12
	v) Grape fruits	8-9	85-90	12
4	Custard apple	5-7	85-90	5-6
5	Dates (ripe)	7-8	85-90	5-6
6	Dates (cured)	0-1	65-70	52
7	Fig, fresh	0-2	85-90	4
8	Fig, dry	0-2	65-70	52
9	Grapes	0-2	85-90	6-8
10	Guava	9-10	85-90	3
11	Jack fruit	11-13	85-90	6
12	Litchi	0-2	85-90	10
13	Mangoes	12-13	85-90	4-7
14	Mangosteen	5-7	85-90	6-7
15	Papaya	9-10	85-90	1-2
16	Passion fruits	7-8	85-90	4-5
17	Pineapple, Mature green	11-13	85-90	3-4
18	Pineapple, ripe	8-9	85-90	4-6
19	Pomegranate	0-2	80-85	4-6
20	Sapota	11-12	80-85	6-8
	Temperature fruits			
21	Apple	0-2	85-90	16-32
22	Apricot	0-2	85-90	2
23	Cherry	0-2	85-90	2-3
24	Peach	0-2	85-90	2-4
25	Pear	0-1	85-90	12-26
26	Persimmon	0-2	85-90	7
27	Plum, prune	0-2	85-90	4-8
28	Quince	0-2	85-90	3-4
29	Strawberry	0-2	85-90	2-3
30	Olive	7-10	85-90	4-6
	Others			
31	Cashew apple	0-2	85-90	4-5
32	Dried fruits	0-2	85-90	41-52
33.	Star gooseberry	0-2	85-90	4-5

Table 8: Maximum acceptable level of water loss in fruits and vegetables during storage

Maximum water loss, %	Commodity
3	Spinach, strawberry, lettuce
4	Carrots bunching with leaves, Sprouting broccolli
5	Apple, tomato, Beans, Beetroots with leaves,
6	French bean, Peas, Green Pepper, black berry, Broad beans, raspberry
7	Beet root, carrots, cauliflower, onions and oranges, parsnip, sweet corn, water cress, potato
8	Asparagus and Brussel sprouts
10	Cabbage, cucumber, celery (white), onions.

iii) **Freezer storage:** It is the low temperature storage method for long term preservation of foods and other biomaterials. This storage is at temperature below 0°C and preferably below -18°C to -40°C (Also discussed in Chapter: Food Freezing Technology). It involves crystallization of most of the water and some of the solute by reducing the product temperature to -18°C or lower upto -40°C. At this temperature the growth of micro-organisms is retarded and enzymes activities are reduced to minimum. The freezing of the foods is categorized into two types (a) Quick/Fast Freezing (b) Slow Freezing.

a) **Quick Freezing:** It is the process by which the temperature of food is lowered to about -20°C with 30-90 minutes. This temperature is achieved by any one of the following method like (i) placing in contact with the coil through which the refrigerant flows (ii) blast freezing in which cold air is blown across the food (iii) dipping in liquid nitrogen or liquefied gases (CO_2, N_2). It is also called cropreservation i.e. very low temperature (-196°C) preservation / storage (iv) freezant spray (v) plate freezer (single, double or pressure) (vi) still air (Sharp freezer (vii) slush freezer (viii) fluidized bed freezer (ix) liquefied immersion {brine (NaCl), di-chloro-di-floromethane}. By quick freezing, most perishable foods can be preserved for several months.

Quick frozen foods maintain their quality and freshness when they are thawed (brought to room temperature) because only very small ice crystals are formed when foods are frozen in this manner. Many micro-organisms can survive by this treatment and becomes active and spoil the food if it is kept at higher temperature Frozen foods should therefore, always be kept at temperature, below -5°C. Enzymes in certain vegetables can continue to act even after being quick frozen and so such vegetable have to be given a mild heat treatment called blanching (above 80^{0}C) before they are frozen to prevent development of off flavours.

The best way of preserving pure fruit juice is by freezing. Properly frozen juice retains its freshness, colour and aroma for a long time. This method is particularly useful in the case of juices whose flavour is adversely affected by heating. The juice is first de-aerated and the vacuum filled with N_2 gas. It is then transferred into containers which are hermetically sealed and frozen. Moulds are some time not affected by this technique. Juice can be kept in good condition for a long time in frozen form at -12 to -17^{0}C by excluding air. It is defrosted before consumption.

b) **Slow freezing:** It refers to the process where the desired temperature -20°C is achieved with in 3 to 72 hours. This is essentially the type of freezing method utilized in the home freezer. It is also known ass sharp freezing. Quick freezing process gives more advantages than slow freezing if compared for overall quality of product.

Advantages of freezing: (i) Neither add nor remove any component (ii) do not alter the flavor (iii) does not diminish the digestibility or cause a significant loss of nutrient value

Disadvantage: (i) Micro-organisms may be reduce in numbers but not destroy (ii) spore are very resistant (iii) unwrapped foods dehydrate rapidly and cause deterioration of flavor (iv)

Advantages of refrigeration over freezing: (i) it takes less energy to cool a food to just above its freezing point than to freeze it. (ii) Require less insulation and refrigeration capacity (iii) No deterioration of texture and flavour as caused by freezing. (iv) No need of thawing before use (as thawing take as much as 48 hrs).

2. Gas storage: Storage of products under modified or controlled atmosphere where atmospheric composition surrounding the commodity is different from that of normal air (78.08% N_2, 20.95% O_2, 0.0, 0.3% CO_2) which usually involves reduction of oxygen and or elevation of CO_2 conc. This storage reduces ripening, senescence and respiration and thereby increases shelf life of the produce. MA and CA differ only in degree of control and CA is more exact. These storage methods in combination with refrigeration markedly enhance storage life of fruit and vegetables.

i) **Controlled Atmospheric (CA) storage:** Storage of commodities under gaseous atmosphere like gas storage, where the composition of atmosphere is controlled accurately. CA storage is some also called chill storage (-1°C to -4°C). In controlled atmospheric (CA) storage, in addition to controlling the temperature (decrease) and RH (increase), the amount of O_2 is reduced while that of CO_2 is increased. CO_2 composition is increased 2-3% by different machines depending on the commodity. O_2 is reduced upto 5-6%. In such storage apple can be stored upto one year. It is used commercially in world. In CA storage, each fruit and vegetables have its optimum level of O_2. It is not advisable do not stop the respiration because if O_2 supply is stop by 100% then fruit and vegetables start respiring anaerobically and causes spoilage due to fermentation. Apple can be stored for 10 month absolutely fresh in CA storage in advanced countries.

- **Temperature control:** Chill storage requires controlled low temperature which is achieved by taking heat away from the storage area by the principle of latent heat of vaporization. If the states changed from liquid to gas or from gas to liquid without affecting its temperature an amount of heat must be added or removed. Liquids like ammonia and freon (Cl_2 F_2) are used as refrigerant. Liquid refrigerant is circulated through evaporator, within the storage area. The temperature of vaporization of the refrigerant is about 5-6°C below the storage temperature. The heat produced form the respiration of fruit and vegetables vary considerably. Products like green beans, peas, spinach, sweet potato, strawberries have a high response rat even at 0°C.
- **Air and humidity control:** The heat produced due to respiration in the vicinity of food surface in cold storage is to be transferred toward the refrigerator. This is achieved by air circulation. The humidity of the circulated air must be controlled. If it is too moist, moisture become condense on the cold food and mould will grow on food surfaces at the common refrigeration temperature. On the other hand if the air is too dry, there will be desiccation of food. The optimum relative humidity of circulated air depends upon the moisture content of the foods and the ease with which they dry out. Most foods store best at a refrigerated temperature when the RH of air is between 80 and 99%. Dry and granular food store well above 50% RH. When foods are to be chill stored for a long time they are protected by packing to prevent loss of moisture. Plastic sacks or moisture resistant coating are used for this purpose.
- **Gas Control:** Continued biochemical changes of fruit and vegetables even under chill storage result in over ripening, softening and loss in

quality. These changes takes place at the normal composition of air (78%, N2, 21% O2 and 0.03%, CO_2) and they can be arrested by using refrigeration in conjunction with modified atmosphere or controlled atmosphere. By raising CO_2 up to 3% and lowering the O_2 level up to 3%, food can be store for long time, with RH of 80-95%. For each food there is an optimum storage temperature, RH, O_2 and CO_2 level which delay ripening (Table 9). The presence of diplenyl vapours can inhibit mold growth in citrus fruits and use of ethylene gas can speed up the ripening and colour development of citrus fruit and bananas. Storage under controlled atmosphere is influenced by the concentration of O_2, CO_2, ethylene and other gases. Other factors which influence storage are crop spp., cultivars, growing conditions, maturity, quality, temperature, RH, packaging and storage conditions.

Lethality of CA Storage: Atmosphere in CA stores, although support plant life at a low level, but not support mammalian life. Atmosphere modified with liquid N_2 are very dangerous. Low oxygen supply and high CO_2 is also lethal to human being. The main disadvantage of CA storage is the internal CO_2 injury of apple also known as brown heart.

Potential Benefits of CA Storage

1. Increase the shelf life/ storage life of vegetables and fruits.
2. Reduces weight loss, discolouration, and toughening, sprouting, fibre formation.
3. Simple design results in an exceptionally long service life of requirements.
4. There is negligible energy consumption.
5. Help to retain organoleptic properties (flavor, taste, turgidity, etc.).

Essential Feature of CA Storage

1. Low temperature (30 -32°F) is maintained by mechanical refrigeration
2. The storage room is specially constructed gas tight for regular CA storage.
3. In CA storage excess CO_2 is removed by dry hydrated lime.
4. O_2 level 3% and CO_2 level 2-5% may be maintained depending of the commodity
5. The load or unload the commodities, the atmosphere mist be flushed out to raise O_2 up to at least 18% and reduce CO_2 a low level

Factors Taken into Account with CA Storage

1. It is expensive. Therefore, only the best & costly fruit should storage.
2. If there is a choice small fruit should be stored as they can be stored better than larger fruit.
3. Fruit should be stored immediately after harvest.
4. Only one type of crop preferable one variety should be stored in the same room.

Table 9: Recommended CA requirement for important vegetable

Vegetables	Temperature °C	Atmosphere Composition	
		CO_2 %	O_2%
Artichokes	0 (0-5)	2-3	2-3
Snap Bean, green	7 (5-10)	2-3	4-7
Broccoli	0 (0-5)	1-2	5-10
Cabbage	0 (0-5)	2-3	3-6
Chinese cabbage	0 (0-5)	2-0	0
Cucumber	12 (8-12)	1-4	0
Fresh lettuce	0 (0-5)	1-3	0
Mushroom	0 (0-5)	Air	10-15
Okra	10 (7-12)	Air	4-10
Onions	0 (0-15)	0-1	0
Onions (bunching)	0 (0-5)	2-3	0-5
Pepper (bell)	12 (8-12)	2-5	0
Pepper (Chilli)	12 (8-12)	3-5	0-5
Radish (Topped)	0 (0-5)	7-10	5-10
Spinach	0 (0-5)	7-10	5-10
Tomatoes	12 (12-20)	3-5	2-3

ii) **Modified atmospheric storage (MA storage):** This type of storage is possible in plastic base films. This is a system of storage where there is no exact control of gases and atmosphere and it is created by use of different packaging material. Sealed or unsealed bags are used to store apple, pear and banana. A major problem with sealed bags is that the atmosphere inside package depends on the temperature, because the permeability of the film to gases is virtually independent of temperature at which produce is normally handled, whereas respiration is temperature dependent. Therefore sealed bags are risky when temperature varies. To avoid brown spot and other CO_2 injuries, sachets of fresh hydrated lime can be included in the bag to reduce the CO_2 conc. @ 100-200g/10 kg of fruits. Attainment of MA in polythene bags filled with produce can be accelerated by evacuating the bags to between 50-85 kilopascals (380-635 mm of Hg) and then sealing them since the polythene film is permeable to N_2, O_2 and CO_2. The pressure inside bags returns to atmospheric pressure but the initial rapid reduction of O_2 concentration is often useful. MA is more economical than CA storage.

3. Pressure Storage: The pressure storage may be at low pressure (Hypobaric) and high pressure (Hyperbaric) storage.

i) **Hypobaric storage:** Preservation of food (usually fresh fruit and vegetable) under low pressure is known as hypobaric storage. In this method the commodities are kept in vacuum tight and refrigerated container (having strength to protect the negative pressure as outside pressure is more). Low pressure inside the container reduces the O_2 tension and facilitative the removal of ethylene and volatile compounds produced by the products which results in retardation of ripening and senescence by decreasing respiration. The level of O_2 and CO_2 is not controlled in hypobaric storage.

It is storage of produce under low pressure. Hypobaric storage is a form of CA storage in which produce is stored in a partial vaccum i.e. storage of fruit under lesss than atmospheric pressure. Hypobaric storage was reported by Brug and Burg (1973) to retard ripening. Fruit is held at about 0.20-0.50 atm. pressure at 59-73°F. Air is passed through the chamber to remove volatile given off by the fruit. It is also called low pressure storage. It is not used commercially but used academically. Ripening of fruit is retarded by hypobaric storage, due to the reduction in the partial pressure of O_2 and for some fruit also the reduction in ethylene level. A reduction in air pressure to 10 kilopascals (0.1 atm.) is equivalent to reducing the O_2 concentration to about 2% at normal atmospheric pressure. Hypobaric storage is expensive to construct, therefore, storage is limited upto high value produce. It is refrigerated storage under reduced pressure and high humidity. MA does not differ in principle from Ca storage except that the control of gas concentration is less precise in MA storage. Flexible films used for pre-packaging are Cellophane, polyethylene, cellulose acetate, polystyrene films etc. Sealed polyethylene bags retard respiration and transpiration and help in increasing the shelf life and retain better quality of the produce. There is very good future scope for MA storage of fruits and vegetables.

ii) **Hyperbaric storage:** Preservation of foods under high pressure (more than atmosphere, 14.7 lb/inch2) is known as hyperbaric storage. Generally liquid foods like fruit juices are stored by this way under pressure and so, due to the every possibility of tissue break down of fruit and vegetable tissue; they are not stored by this method. Pressure due to inclusion of inert gas (CO_2, N_2 etc.) helps to retain the natural flavor of juice and check micro-organisms deterioration as done for carbonated soft drinks. The level of O_2 and CO_2 is not controlled in hyperbaric storage.

4. Refrigerated transport (Vans): This type of storage is practiced in perishable products that are transported to longer distance/market followed by storage at cold / CA storage or direct sale into the market.

5

Edible Coatings for Shelf Life Extension of Fruits and Vegetables

Loitongbam Sophia Devi

Directorate of Research, Central Agricultural University, Imphal-Manipur

Introduction

Fruits and vegetables are essential constituents of daily diet and are highly demanded in the recent years from most of the population. They are reservoir of vitamins, essential minerals, antioxidants, bio-flavonoids, dietary fibres and flavour compounds which fall easily victim to abiotic and biotic adversities. Fruits and vegetables are highly perishable and during the post- harvest, there are considerable losses due to microbes, insects, respiration and transpiration.

The important quality factors of fresh produce contributing to the marketability are texture, colour, appearance, flavour, nutritional value and microbial safety. These quality factors are measured by plant variety, ripening stage, maturity stage pre-harvest and post-harvest conditions. The Post-harvest losses of fruits and vegetables are a serious problem because it rapidly deteriorates them during handling, transport and storage. Edible coating over fruits and vegetables are used to improve their quality and shelf life.

Edible coatings are used for extension of shelf life of fruits and vegetables. These can also be safely eaten as part of the product and do not add unfavourable properties to the foodstuff. Edible coatings or films increase the shelf life of fruits and vegetables and are environment friendly. In recent years, new edible films and coatings have been developed with the addition of various and edible herbs, antimicrobial compounds to preserve fresh fruits and vegetables. Edible coatings to prevent loss of firmness and moisture. They control maturation, development and respiratory rate. Edible coatings prevent oxidative browning and decrease growth of microorganism in fruits and vegetables for example, tomato, cucumber, and cherries etc. Edible coatings or edible films are contributed to enhance the shelf life of fruits and vegetables by reduction of moisture loss solute migration and gas exchange etc.; as well as by reducing the physiological disorders. Edible coatings have high potential to control browning, discolour activity, off flavour, microbial activity of fruits and vegetables and to extend shelf life.

Edible Coating

History

Edible coatings or edible films have been used for centuries in the food industry to preserve food products. Edible coatings have been used since 12th century in China. It was not until 1922 the waxing on fruits was invented and first time was commercially applied on fruits and vegetables. Edible films and coatings form a barrier for chemical, physical and biological changes. At the time of purchasing fruits and vegetables, consumer judge the freshness and quality of the produce on the basis of its appearance . The most common and challenging problem are to maintain and control fresh quality, growth of spoilage and pathogenic microorganism in fresh cut fruit industry. The solution of this problem is edible coating . Edible coating provides an additional protective coating for fresh fruits and vegetables and can also provide the same effect as modified atmosphere storage in modifying internal gas composition. Recently, various edible coatings were applied successfully for preserving fruits and vegetables such as orange, apples, grapefruit, cherries, cucumber, strawberry, tomato and capsicum were applied successfully. Edible coating of fruits and vegetables is successful or not totally depends on the control of internal gas composition.

Definition

Edible films and coatings are defined as any thin material used for wrapping or coating food to extend shelf life of the product which may be consumed together or removed before consumption. They can improve quality of food products by protecting them from physical, chemical and microbiological deteriorations such as moisture loss, enzymatic browning reactions, microbial spoilage and lipid oxidation. The thin layer which is created by edible materials to cover the foodstuff is called edible coating, shaped previously and can be placed on or between food ingredients is called edible film. Edible coatings are in liquid form and applied to the product by plunging it into the solution. Edible films are shaped like solid sheet and then applied by wrapping around the product.

Different type of materials were used for coating and wrapping various fruits and vegetables to extend their shelf life, and this is eaten together with foods, with or without removal is considered an edible coating. Edible coating or edible films provide shiny appearance to fruits and vegetables. Thickness of edible coating is generally less than 0·3 mm. The main characteristic feature of edible coating is to increase shelf life of fresh or processed fruits and vegetables and it is protected from postharvest damages and environmental damages. An edible coating protects outer membrane of fresh fruits and vegetables . The edible coatings are served as carrier of texture enhancer, antioxidants and it is used as a nutraceutical. Under high relative humidity, edible coating should be stable, generally recognised as safe. Edible coating or edible films are mostly tasteless, colourless and odourless they should have good mechanical properties. Edible coatings have good gas barrier and moisture barrier properties.

Edible Coating Applied on Different Fruits and Vegetables

Fresh fruits and vegetables are highly perishable and approximately 50% fresh produce are deteriorated during harvest, handling, transportation and storage. Edible coatings play a very important role to handle this situation. Edible Coatings are applied on whole and fresh-cut fruits and vegetables. Fruits and vegetables which has been coated are-

Fruits: Edible coated fruits are Orange, Apple, Grapefruit, Cherry, Papaya, Lemon, Strawberry, Mango, Peach etc. and fresh-cut Apple, fresh-cut Peach, fresh-cut Pear etc.

Vegetables: Tomato. Cucumber, Capsicum, Cantaloupe and minimally processed Carrot, fresh-cut Potato, fresh-cut Cabbage, fresh-cut Tomato slices, fresh-cut Onion, Lettuce.

Fig. 1: Edible coated fruits and vegetables.

Classification of edible coatings

Edible coatings are having hydrophobic group, for example lipid-based or waxes, and hydrocolloids or hydrophilic group, for example polysaccharides-based, protein-based or combination of both groups to improve function of edible coating. Edible Coatings are not chemically synthesised and it is natural. It is generally used for good appearance and preservation of fruits and vegetables. The main advantages of edible coating are its edibility, non-toxic nature and cost effective as compared to other synthetic coating. Edible coating materials are generally made up of polysaccharides, proteins and lipids. The edible coatings are mainly divided into three classes; these are following:

- Hydrocolloids: e.g., polysaccharides, proteins and alginate.
- Lipids: e.g., fatty acids, acryl glycerides and waxes.
- Composites: e.g., protein/protein, polysaccharides/protein, lipid/polysaccharides

Different types of edible coating

Hydrocolloids: Hydrocolloids are originated from animals, vegetables, microbial or synthetic, they are hydrophilic polymers. They have hydroxyl group and may be poly-electrolytes such as Alginate, Carrageenan, Pectin, Carboxy Methyl Cellulose, Xanthan gum and Gum Arabic. Today, hydrocolloids are used in wide range as a coating forming solution to coat and control the colour, texture, flavour and shelf life of fruits and vegetables. The hydrocolloids are divided into two classes:

- **Polysaccharide-based:** The most common polysaccharides used for edible coating of fruits and vegetables are chitosan, starch, alginate, cellulose, gum etc. Polysaccharides based edible coatings having poor moisture barrier properties, it is water soluble. But it contain moderately low O_2 permeability. Mostly, polysaccharides based edible coatings are applied on fresh and minimally processed fruits and vegetables, by creation of modified atmosphere condition to reduce their respiration rate. Polysaccharides give crispiness, hardness, compactness, thickening quality, adhesiveness and viscosity to a variety of edible coatings. Polysaccharides are made up of polymer chain, having excellent gas barrier properties, resulting in desirable modified atmosphere that extend the shelf life of the fruits and vegetables without forming anaerobic conditions. These coatings are given below-

Cellulose Derivative

Cellulose is easily found in nature. It is made from long chain of anhydro-glucose polymer. Cellulose is reacted with alkali, then it is treated with appropriate chemical reagent and then it forms substitute of cellulose chain of anhydro-glucose monomers. This process has been done under controlled conditions . The substitution reagents are given below-

a) Methyl cellulose- formed by chloromethane,

b) Hydroxypropyl cellulose (HPC)- substitute reagent is propylene oxide,

c) Hydroxypropyl Methyl Cellulose (HPMC)- mixed substitution of chloromethane & propyl oxide,

d) Methyl ethyl Cellulose (MEC)- mixed substitution of chloromethane & chloroethane,

e) Carboxy Methyl Cellulose (CMC) - monochloroacetic acid.

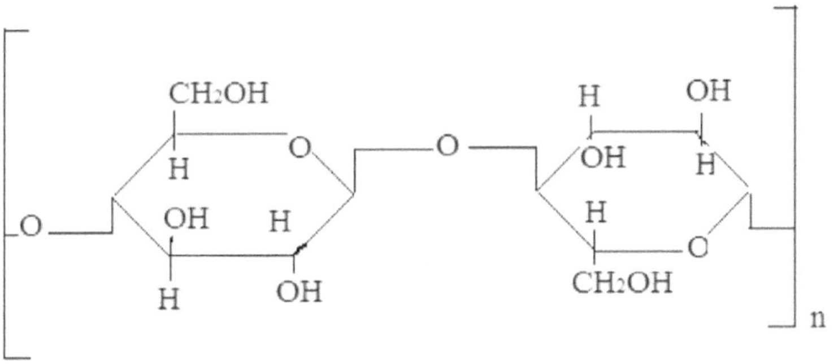

Fig. 2: Structure of Cellulose

Cellulose derivatives are made from polymer chain of two repeating unit of anhydro-glucose or β-glucopyranose residue attached through 1, 4-glucosidic linkage. In this chemical structure, 'n' indicates the no. of glucose monomers of polymerisation. The DS (Degree of substitution) are defined as, the average no. of –OH group substituted per anhydro-glucose unit. Generally the modified cellulose derivative edible coatings and films give colourless clear solution, odourless and tasteless. It is having good resistant property of oil and fats and soluble in water and partially permeable to moisture and gas transmission . The mechanical barrier property of cellulose edible coating and films are based on molecular weight of cellulose, that is the higher molecular weight of cellulose the excellent barrier properties . Methyl Cellulose has good moisture barrier property as compared to others and it is lowest hydrophilic.

These modified cellulose derivatives are found in granule and powder form; it is hygroscopic in nature. However, cellulose derivative coatings and films have poor gas and water barrier properties. Polysaccharides derivatives are hydrophilic in nature that's why they have poor mechanical properties.

HPMC: Hydroxypropyl Methyl Cellulose is commonly used for edible coating. It is non-ionic water soluble macromolecule, is capable to form gel on heating. HPMC is by product of propyl oxide and chloromethyl.

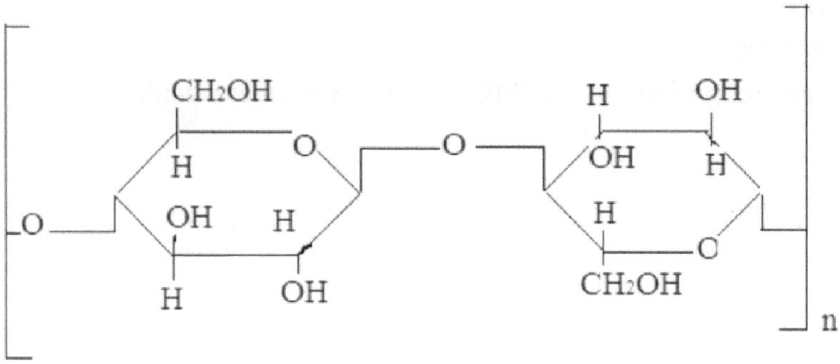

Fig. 3: Structure of HPMC

The Hydroxypropyl methyl cellulose is divided in two regions, one region contain Methoxy group and this region is hydrophobic in nature, and other region have HydroxyPropyl group that are hydrophilic in nature. HPMC are soluble in cold water and after dissolve in water it gives different viscosity characteristics solution, it's totally depend on their degree of polymerization and substitution. The different types of HPMC are found in market, which contain different viscosity and different molecular weight.

Starch Derivatives

Starch is the most common polysaccharide. Starches are obtained from cereal grains such as wheat, maize, rice etc., and cereal grains are belong to Poaceae family. Potato and other tubers, legumes are also good source of starch. Starches are composed of anhydro-glucose units. Starch is a homopolymer composed of polysaccharides components- one is water soluble amylose and second is water insoluble amylo-pectin. On the other hand the amylose are a linear or unbranched chain of D-glucose (1-4 linkage) and the amylopectin is a branched chain of glucose (1-6 linkage). Amylose are good starch derivatives for film and coating material coatings and films of high amylose corn starch or potato starch are contained high stability during long period preservation. These chains are held by the help of glucosidic linkage. Starch derivatives are breakdown by the help of amylase and the final product is dextrin. Starch amylose for example corn starch is excellent source for coating and film production. Corn starch based films and coatings demonstrated of physical characteristics resemble to plastic films and coatings, they are tasteless, odourless and colourless and non-toxic. It is biologically absorbent, semi-permeable to oxygen and carbon-dioxide.

Chitosan

Chitosan is derived from chitin, it is an edible polymer. Chitin is mainly found in crustacean animal shells. Chitosan is the most common non-toxic and natural product after cellulose for formation of edible coating.

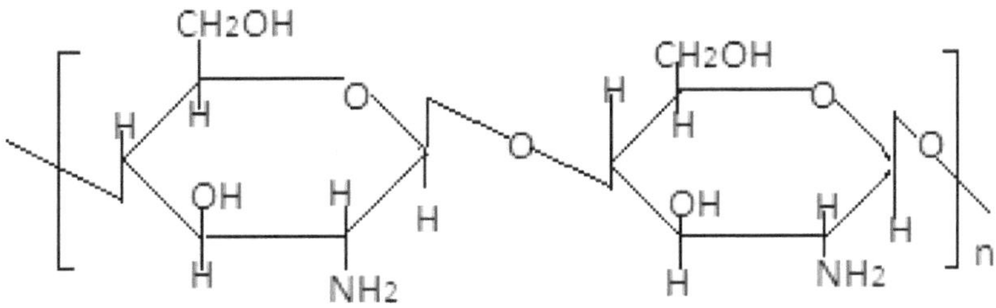

Fig. 4: Structure of Chitosan

Gums

Mostly all gums are polysaccharides it consists of sugars. Gums are used for preparation of edible coating on fruits and vegetables because of its texture capability. Generally, gums are divided into three parts-

a) Exudate gums (e.g., gum Arabic)

b) Extractive gums (e.g., guar gum)

c) Microbial fermentation gums (e.g., xanthan).

Gums are soluble in water. Xanthan gums are prepared by microbial extraction by fermentation process. It is rapidly spread in water due to this high viscosity are found readily in cold and hot both stages. A mixture of gum Arabic, guar gum and xanthan gum are used to form equal coatings with good adhesion property in wet batters.

Alginate

Alginate is obtained from brown seaweed, which is related to the Pheophyceae family. Alginate contains salts of Alginic acid. Alginate having a linear chain copolymer of D-mannuronic acid L-guluronic acid monomers, the chemical formula of alginate is $(C_6H_8O_6)$ n. It is found in white, yellow, fibrous powder form. Alginate is commonly used in form of Sodium alginate this is extracted from brown algae. Alginate contain excellent barrier for moisture and water vapour.

Alginate maintains good properties useful in food application. It has a unique colloidal property, which contains stabilising, thickening, suspending coating or film producing gel forming and stabilising emulsion.

Pectin

Pectin is a group of polysaccharides which is derived from plant; it is naturally found in fruits and vegetables. Pectin is good for low moisture fruits and vegetables but is not a good moisture barrier. It is a heterogeneous group of acidic polysaccharides. It is commonly found in peel of citrus fruits and apple pomace. The structure of pectin polysaccharide are very complex, are made up of β-1, 4-liked-galactouronic acid residues. Pectin is a complex polysaccharide; it composed from 1, 4- linked d-galactouronic acid residue.

Agar

Agar gum are obtained from red seaweed Rhodophyceae family, it is a galactose polymer. Agar is hydrophilic colloidal in nature it contains a mixture of agro pectin and agrose. It is used widely in microbiological media to give firmness. Agar forms a strong gel melting point for above the initial gelation temperature. Agar gels are set on cold temperature and melt on high temperature. Because of their capacity to form hard gel at low concentration and of its extraction. Agar is widely used in food industry.

Protein-Based Edible Coating

Protein based edible coatings are derived from animals and plants. The plant based protein edible coating material are milk protein casein, whey protein, zein (from maize), gluten (from wheat), soy protein etc. and the animal based protein are egg albumen, collagen etc. Protein based edible coating consist excellent barrier properties for aroma, oil and oxygen and it gives strength but it is not effective barrier for moisture. The reason of its excellent barrier property for oxygen is its tightly packed hydrogen bonded structure. Protein based edible coating has good O_2 barrier property at low relative humidity. Protein based coating are not good barrier for water vapour due to its hydrophilic nature but it consist good organoleptic and mechanical properties.

Lipid Based Edible Coating

The lipid based edible coatings are used from many years for preservation of fruits and vegetables. They provide shiny and glossy appearance to food. Most common lipid based coating materials are carnauba wax, bees wax, paraffin wax, and mineral or vegetable oil. Lipids are having good water barrier capacity. Wax coatings contain very good moisture barrier properties as compare to other lipid based coating and non-lipid coating. Oil, fat and wax based coatings are not easily applied to the surface of fruits and vegetables because of its greasiness and thickness and it gives rancid flavour. Combination of lipid and polysaccharides, protein are used in coating material improve their barrier properties. Most common lipid based coating materials are as follows-

- Waxes
- Lacs
- Fatty acids and alcohols
- Acetylated glycerides
- Cocoa-based material

Composites Based Edible Coating

Composites or multicomponent films and coatings contain combination of protein, polysaccharides and lipid based material. This is used to enhance and improve mechanical strength, moisture and gas barrier properties of edible coatings and films

The bilayer composites consist two layers combined with same or different coating materials such as protein/protein, polysaccharides/protein, lipid/lipid, lipid/polysaccharides etc. In this type of coating materials include Sucrose, propylene, monoglycerides, proteins, water, waxes and fatty acids.

Sl. No	Fruit and vegetables	Used Edible coating
1.	Fresh-cut Apple	Sodium Alginate, Gellan Gum, Sunflower oil. Whey protein concentrate, WPI & WPP.
2.	Blueberry	CMC, Chitosan, Monoglycerides, Sodium alginate, Calcium casienate.
3.	Tomato	Aloe vera gel
4.	Grapes	Aloe vera gel.
5.	Mango	Chitosan, Aloe vera. Tapico flour, Sago flour, Soy protein. Chitosan.
6.	Apple	Neem oil, Marigold flower extract, Guar gum & Aloe vera. Soybean gum, Paraffin wax, Jojoba oil & Arabic gum. WPC, SPI, Alginate, Carrageenan.
7.	Fresh-cut Pineapple	Pectin & Alginate.
8.	Strawberry	Sodium Alginate & Calcium Alginate gel. Arabic gum & Arjan gum Psyllium mucilage. Chitosan & Tragacenth. Linseed mucilage extract, Chitosan. Pectin, PVA, Starch, Soy protein, Gluten.
9.	Cantaloupe	Aloe vera gel. Pectin, Chitosan.
10.	Banana	PVA, CMC, Tannin.
11.	Guava (fresh-cut)	Aloe vera juice.

13.	Pistachio	Chitosan and Aloe vera gel.
14.	Orange, Tomato, Mango, Papaya, Guava, Mushroom.	Veg. oil, Cellulose gum, Emulsifier.
15.	Potato	Chitosan, WPC, Coconut oil.
16.	Papaya	Chitosan, Aloe vera, Papaya leaf extract.
17.	Plum	CMC, Pectin.
18.	Sapota	Aloe vera juice,
19.	Pumpkin, Carrot, Radish, Cantaloupe, Cucumber.	Chitosan, calcium salt.
20.	Cucumber	Gum Arabic Powder. Beeswax, High Methoxy Pectin,
		Sorbitol, Emulsifier. Chitosan, Aloe vera. Paraffin wax. Sodium Alginate, Starch, Glucose, Sucrose, Chitosan, Sorbitol, PVA.
21.	Apple, Potato, Cucumber.	Potato starch, Guar gum, Pea starch.

Applying Methods of Edible Coating

Edible coatings should be applied on fruits and vegetables by different methods. These methods are:

- Dipping
- Brushing
- Extrusion
- Spraying
- Solvent casting

The dipping method is used widely for applying edible coatings on fruits and vegetables, in this method Fruits and Vegetables are dipped in coating solution for 5-30 sec. It is easy to apply on mostly fruits. While Brushing method gives good result, Edible Coatings applied on generally, Beans and highly perishable Fruits and Vegetables such as strawberry, berries. Other three methods spraying, extrusion and solvent castings are also used in food industry. Extrusion method depends on thermoplastic properties of edible coatings; it is best technique for applying of EC for industrial purpose as compared to other methods.

Advantages

Advantage of edible coatings are included

- Edible coatings improve retention of acids, colour, flavour and sugar.
- Maintain quality of fruits and vegetables during storage.
- Reduce weight loss and firmness loss.
- Decrease polymer packaging and waste.
- Edible coatings can be consumed along with fruits and vegetables, they contain health beneficial nutrients.

Disadvantages

Edible coatings have some disadvantages.

- Thick coating can prohibit Oxygen exchange, causes off-flavour development.
- Edible coatings have good gas barrier properties which causes anaerobic respiration due to this normal ripening process is disturbed in fruits and vegetables.
- Some edible coatings are hygroscopic in nature, which helps to increase microbial growth.

Herbal Edible Coatings: A New Concept

Herbal edible films is a budding technique in food engineering. This kind of edible coatings can be synthesized using extracts from Garlic, Rosemary, Cinnamon, Turmeric, Ginger, Tulsi, Neem and Aloe vera, etc. These herbal extracts have already proven for its antimicrobial, antioxidant and therapeutic properties and can also act as nutraceuticals.

1. Aloe vera

Aloe vera gel is extensively used in edible coating industries because of its antimicrobial properties. Extracts from ginger, turmeric, mint, clove bud, etc are also using in edible coating preparations. Aloe vera is a semi-tropical plant with good medicinal properties. "Aloin and Aloe-emodin" are the major components of Aloe vera extract. This extract has anti-fungal, anti-bacterial, anti-inflammatory properties. It also consists of components like glycoproteins, polysaccharides, salicylic acids, phenolic compounds, lignins, amino acids, vitamins, saponins and enzymes. It can suppress respiration and can reduce microbial growth, thus can enhance the shelf life of fruits and vegetables. It could retain the volatile flavor components in fruits and vegetables and thus could prevent softening and oxidative browning.

2. Neem

Neem is considered as the holiest, therapeutic and a non-lethal plant, which has antimicrobial properties. The antibacterial properties of neem act against pathogens like *Salmonella, Staphylococcus, E. coli,* Vibrio and other microbes. Nimbidin and Nimoid are the major components of neem extract. These components can act as an antimicrobial operator. Neem extricate is utilized in numerous biodegradable consumable covering to improve realistic usability.

3. Tulsi

Tulsi is the most sacred homegrown plant of India and is likewise called "Sovereign of Herbs. Tulsi a member of the Ocimum family with botanical name "*Ocimum sanctum*" and known as "*Ocimum tenuiflorum*". Tulsi has numerous components such as alkaloids, tannins, saponins, glycosides, and fragrant mixes. It additionally improves shelf life. The extract and oil of Tulsi had numerous helpful properties, for example, antibacterial, antifungal; antiviral, antioxidative and go about as antioxidative agents too.

4. Ginger

The ginger plant belongs to the Zingiberaceae family. Ginger consists of 80.9% water, 12.3% carbohydrates, 2.4% fibre, 2.3% proteins, 1.2% minerals, and 0.9% lipids. It contains properties and antiviral property. It can act as an antioxidant because of the presence of phenolic compounds. It reduces the transpiration rate and also inhibits shrinkage. The components like β-sesquiphellandrene, cis-caryophillene, Zingiberene,α-farnesene, α-β-Bisabolene present in zinger are responsible for the antimicrobial activity of ginger.

5. Turmeric

The turmeric is used in food coating industries because of its antioxidant, antiviral and antifungal properties. It lowers the transpiration rate and increases the quality and thus the shelf life of the vegetables. The colour of the turmeric is produced by curcuminoid. This protects the coating materials from oxidative destruction. Turmeric oil can delay the enzymatic browning process and also proven for its antioxidant properties.

6. Garlic

Garlic prevents the oxidation reaction and thereby imparts a high life span for fruits and vegetables. It can also act as an antiinfective agent. Garlic has an antimicrobial activity due to the presence of phenolic compounds and organosulfur compounds and also used to treat many microbial diseases.

7. Cinnamon

The cinnamon is rich in cinnamaldehyde and β - Caryophyllene and other terpenes. Cinnamaldehyde is the major constituent that gives the flavour and thus used as food additives. It has antimicrobial activity due to the presence of compounds like cinnamaldehyde and eugenol which present naturally in the cinnamon. Thus cinnamon is used in the making of edible films.

8. Rosemary

Essential oils from rosemary (Rosmarinus officinalis L.) can be used in edible film preparation. This essential oil has natural antioxidant and antimicrobial properties. It imparts towards the food safety and quality of fruits and vegetables. It lowers the lipid oxidation and oxygen transfer rate and also increases the water vapour barrier.

9. Lac-based formulation (fresh coat) technology for fruit coating application at commercial plants

A lac-based formulation was developed for fruit coating applications on apple and citrus fruits like kinnow and orange. The formulation yielded good results in respect of gloss and firmness to kinnow. To evaluate and assess the performance of fresh coat on kinnow fruit at commercial level, about 40 litres of the formulation was tried at 7 kinnow waxing and grading plants at Abohar (Punjab). Approximately 83 tonnes of kinnow fruits were coated with the above formulation. There are 17 kinnow waxing and grading plants at Abohar in Punjab, and they use commercial formulations such as NU Coat Flo, Citrashine, Stay fresh 451 and Stayfresh high shine for coating kinnow fruits. Chemical fungicides are added to control post harvest diseases in stored kinnow fruits. Application of fresh coat produced better results in respect of gloss, spread area and firmness to the fruits as compared to commercial wax. Other advantages of lac-based formulation over the available commercial ones observed are: it is natural, nontoxic and composition of lac-based formulation is well specified, no fungicide addition is required since lac has inherent property of antifungal properties, lac-based formulation does not produce any obnoxious smell during waxing operation, it is water-based and thus making the plant more versatile.

Conclusion

The edible coating is the most efficient method for enhancing the postharvest quality of fruits and vegetables. The edible coating is an effective technique to preserve fruits and vegetables. It does not create any harm to the environment and can act as an oxygen barrier, moisture barrier thus delaying the fruits from ripening. The herbal coating is the new and developing technology that has antioxidant, antimicrobial and antifungal properties. Since the herbal extract consists of vitamins and minerals it can also impart towards the nutritional value of edible coatings. Probably the herbal edible coating will be the best alternative method for the wax and chemical coating in fruits and vegetables. Herbal edible coated fruits and vegetables contained nutrients and act as medicines.

References

Kohli, D., Srivastava, A., Kumar, S., Upadhyay, S. and Kaur, G.(2018). Effect of edible coating on quality of fresh cut sliced papaya. International Journal of Food Science and Nutrition. Volume 3, Issue 2, pp. 64-66.

Patel, C. and Panigrahi, J. (2019). Starch glucose coating-induced postharvest shelf-life extension of cucumber. Food Chemistry, Vol. 288, pp.208-214.

S. Chauhan, K.C. Gupta & M. Agrawal (2014). Development of A. vera gel to control postharvest decay and longer shelf life of Grapes, Int. J. Curr. Microbio. App. Sc. 3(3), pp. 632-642

S. Chauhan, K.C. Gupta & M. Agrawal (2014). Efficacy of chitosan and calcium chloride on postharvest storage period of Mango with the application of Hurdle Tech. Int. J. Curr. Microbio. App. Sc. 3(5), pp. 731-740

S. Chauhan, K.C. Gupta & M. Agrawal (2014). Efficacy of natural extract on the storage quality of Apple, Int. J. Curr. Microbio. App. Sc. 3(3), pp. 706-711

6

Minimal Processing of Fruits and Vegetables

Alok Nath

College of Horticulture, SDAU, Jagudan, Mehsana-384460, Gujarat

Introduction

In this modern era, educated and health-conscious people are increasingly demanding such food products that preserve their nutritional value, retain a natural and fresh colour, flavour and texture, and free from chemical preservatives. Besides, urbanization has extensively affected the life-style of people all over the world. Competition has made their life extremely busy and time has become a very important factor. Hence, recently there has been increase in demand for ready-to-cook or ready-to-use fresh food products of fruits and vegetables along with convenience and safety. This new demand by consumers has opened a new avenue of food preservation known as minimal processing. The products prepared by this processing method are known as fresh cut or pre-pared products. Minimally processed products are also known as lightly processed, partially processed, fresh processed, fresh-cut, and pre prepared. It can be defined in many ways for example as "least possible treatment to achieve a purpose". A more specific definition is as the method that "preserve foods but also retain to a greater extent their nutritional quality and sensory characteristics by reducing the reliance on heat as the main preservative action. In a broader way it can also be defined as the processed foods have been slightly altered for the main purpose of preservation but which does not substantially change the nutritional content of the food. Generally minimal processing involves cleaning, washing, trimming, coring, slicing, shredding before packaging and storage at specific temperature. Some of the examples of minimally processed fruits and vegetables include sliced mango cheeks/melons/papaya/apple, pieces of pomelo segments, peeled and cored pineapple, chilled peach, pomegranate arils dressing, shredded lettuce, shelled peas, cauliflower florets, peeled sliced potatoes, diced onion etc. Minimally processed products are important to food service industry such as eateries and catering companies as they offer many advantages over traditional products, with respect to convenience, expense, labor, and hygiene. Even if it is very much liked by people, the production of minimal processed products is limited due to rapid deterioration and short shelf life. Whereas most food processing methods stabilize the products and prolong their shelf-life, minimal processing of fruits and vegetables increases their perishability.

Physiology and Shelf Life of Minimally Processed Fruits & Vegetables

Fruits and vegetables undergo a number of physiological changes while they are intact to their plants. These changes become more prominent when fruits and vegetables go under knife during minimal processing i.e. peeling, slicing, abrasion, shredding etc. The physiology of minimally processed fruits and vegetables is essentially the physiology of wounded tissue. Minimally processed fruits and vegetables have very high respiration rates as compared with the intact, unprocessed or raw ones. The rise in respiration rate indicates a more active metabolism and usually a faster rate of deterioration. This leads to rapid loss of acids, sugars and other components that determine flavour and nutritive quality values. Suitable packaging materials are required to store these products as there is higher rate of oxygen demand. The packaging material must have sufficient permeability to gases to prevent fermentation and development of undesirable changes in flavour and odour. During minimal processing the plant tissues are wounded and exposed to stress condition that leads to increase in respiration, ethylene production and other biochemical changes like change in colour, flavour, texture and nutritional quality (sugar, acids and vitamin content). Besides, microbial proliferation may be accelerated simultaneously.

Respiration

Fruits and vegetables have live cells that can respire even after harvesting. During respiration sugars, organic acids are oxidized to CO_2, H_2O and energy. The released energy is used in the synthesis of proteins and carbohydrates that constitute plant tissues. After harvesting, the supply of energy to the fresh cut fruits or vegetables is stopped even though many of reactions inside them still go with a limited supply of energy. The rate of respiration of harvested products is proportional to their rate of respiration. Higher the rate of respiration the shorter the shelf life of these products. The respiration rate of peeled and sliced ripe kiwifruit is double of the whole fruit, but ripe bananas are unaffected by peeling and slicing. The respiration rates of iceberg and lettuce cut as salad pieces (2 to 3cm by 2 to 3cm or 0.75 to 1.18 inches by 0.75 to 1.18 inches are 20 to 40% higher than rates of the respective intact heads. In case of shredded lettuce and shredded cabbage the respiration rates are 200 to 300% greater than those of the intact heads and remain high throughout the storage period.

Oxidative Browning

Oxidation reactions occur in fruits and vegetables when they are cut and their tissues are exposed to atmosphere. The broken cells allow substrates and oxidase enzymes to come in contact with each other. Oxygen in the air can cause sliced fruit to brown, a process called enzymic browning (an oxidation reaction). Phenols and the enzyme phenolase are found in the cells of the apple, and when these are exposed to oxygen in the air, for example through slicing, the oxygen causes a reaction. The phenolase changes the phenols into melanin, which has a brown colour. To stop the oxidative reaction, the phenolase enzymes need to be denatured. Oxidative browning at the cut surface is the controlling factor in storage of various minimally processed fruits and vegetables. Phenylalanineammonialyase (PAL) is a key enzyme in the synthesis of phenolic compounds. It catalyzes the first reaction in the biosynthesis of plant phenylpropanoid products. The phenolic compounds can be oxidized by polyphenoloxidase, producing brown polymers leading to tissue browning in lettuce. Different fruits and vegetables take different time intervals to turn brown. For example, when apple or banana are cut or sliced, the cut surface usually turns brown within an hour while vegetables like lettuce takes several hours to turn brown.

Ethylene

Ethylene is a plant hormone that acts in concert with other plant hormones (auxins, gibberellins, kinins and abscisic acid) to exercise control over the fruit ripening process. Low concentration of 0.1-1.0 microlitres is sufficient to trigger the ripening process in climacteric fruits. It has autocatalytic activity because of which such small quantities can trigger further release of large quantities of ethylene by the fruit tissue. Very little response is only seen to exogenous application of ethylene in case of non-climacteric fruits. It has various effects on the growth, development and shelf-life of horticultural produce. It is more or less produced by all the parts of the higher plants such as leaves, roots, flowers, fruits, tubers, seedlings etc. There are two systems for the regulation of ethylene biosynthesis. Ethylene may be produced from methionine via a intermediates S-adenosylmethionine (SAM) and 1-aminocyclopropane-1-carboxylic acid (ACC). The conversion of SAM to ACC by the enzymes ACC Synthase(ACS). In higher plants, ACC can be removed by conjugation to form malonyl ACC (MACC) or glutamyl ACC (GACC). Ethylene forming enzyme (EFE) or ACC oxidase (ACO) is required to convert ACC to ethylene. ACO is a liable enzyme and sensitive to oxygen and attached to outer layer of the plasmalemma. Factor that effect the activity of the ACS includes fruit ripening, senescence, auxin, physical injuries and chilling injury. This enzyme

(ACS) is strongly inhibited by aminooxyacetic acid(AOA), rhizobitoxine and amino ethoxy vinayl glycine (AVG). ACO is inhibited by anaerobiosis, temperature above 35°C and cobalt ions. Among various chemical used for extension of shelf life of fruits 1-MCP has been found to be very effective. Many fruits respond to 1-MCP in extension of storage life by retarding the process of ripening. Wounding plant tissues accelerate ethylene production rate. Ethylene produced by the minimal processing has been found adequate to hasten the softening of banana and kiwifruit, and chlorophyll loss in spinach. It has also been found that the ethylene level increases in proportion to the severity of injury in many fruits and vegetables.

Loss of Nutrients

Various fruits and vegetables contain a number of nutrients like vitamin C. During minimal processing of fruits and vegetables undergo different processing operations such as handling, storage and primary processing. The vitamin C content of sliced, cut or bruised fruits or vegetables may decrease hastily. There are two aspects of ascorbic acid degradation. First, ascorbic acid can be oxidized due to mechanical damage as a result of cutting. When tissues of fresh fruit or vegetable are wounded during slicing, cutting, chemical reactions are initiated, which shorten the shelf life of the product.

Loss of Moisture

Most of the fruits and vegetables contain 70-80% moisture while some vegetables like leafy vegetables and melons contain almost 92-95% moisture. The tubers crops like cassava, yam and corms contain less moisture (around 50%) and are more starchy. Moisture plays an important role in fruits and vegetables because many of the nutrients exist in soluble state in them. The tissues of fruits or vegetables are in equilibrium with an atmosphere at the sane temperature with relative humidity of 99%-99.5%. Moisture is lost when there is reduction of water vapour pressure in the ambient atmosphere around the tissues. In the whole fruit or vegetable, moisture in intercellular space is not directly exposed to the outside environment. When a fruit or vegetable is cut, it gets exposed to the outside environment and starts to lose moisture rapidly. Hence, to maintain the lowest possible vapour pressure deficit, minimally processed products are wrapped in semi permeable films. The higher moisture content makes the fruits, vegetables and flowers perishable as it is easily vulnerable to attack by microorganisms. Further moisture is lost during the biological activity of these commodities which deteriorates its quality in terms of freshness. Therefore, retention of the moisture or prevention of loss of moisture is one of the important considerations in planning a storage technique or strategy for extension of shelf life in minimal processing.

Methods of Minimal Processing

Minimal processing is generally considered as a processing that does not include thermal treatments. However, now a day some scientists contrast it with thermal processing where there is least loss in quality of fruits or vegetables. Thus, we can categorize minimal processing methods in two groups i.e. i) Minimal processing with non-thermal methods ii) Minimal processing with thermal methods.

Minimal Processing with Non-Thermal Methods

Long shelf-life in foods has been associated with thermal processing, solely or in combination with other methods of preservation. Thermal processing diminishes the freshness as well as nutritional quality of fruit and vegetables. Hence, non–thermal methods are highly preferred in case of minimal processing of fruits and vegetables. There are various non-thermal methods for minimal processing which have been listed as follows:

- Irradiation
- High pressure processing
- Pulsed electric field
- Ultra sound
- Ultraviolet radiation

Among all these methods, only high pressure processing and irradiation have been well developed and commercialized.

Minimal Processing with Thermal Methods

Thermal methods are widely used for the preservation and preparation of fruits and vegetables. It brings about desirable changes such as protein coagulation, starch swelling, textural softening and formation of aroma components. But, it also leads to some undesirable changes such as loss of heat sensitive nutrients, formation of thermal reaction components of biopolymers, loss of freshness, flavor and texture. Some of the important thermal methods for minimal processing are as follows:

- HTST processing with conventional processes
- Aseptic processing
- Semi-aseptic processes
- Sous-vide processing (cooking of packaged raw food under vacuum)
- Infrared heating

Minimal Processing in Practice

Minimal processing of fruits and vegetables may be carried out at the source of production or at regional and local processors. However, it also depends on the perishability of the fresh cut produce. Generally, the minimal processing of fruits or vegetables comprises of various unit operations such as harvesting, procurement and assessment of quality of raw material, cleaning, washing, peeling, slicing, packaging and storage.

1. Harvesting

Harvesting of fruits or vegetables for minimal processing is the foremost unit operation. Cultivation of the produce for minimal processing should specify cultivars and cultivation techniques including sowing time, pesticide, fertilization application, and harvest conditions. The harvesting operation should best be carried out in the morning hours due to cooler atmospheric temperature. After harvesting the produce should be pre-cooled to 1°C to extend the shelf-life. Afterwards, it is processed at earliest to reduce various undesirable changes in it.

2. Raw Materials

The raw material for minimal processing should be of good quality or class. All the types of raw material are not appropriate for this type of processing. The study of various cultivar varieties of different vegetables showed that not all varieties of the specified vegetable can be used for the production of processed vegetables. Different varieties have different yield, rate of respiration, sensitivity to physiological disorders and diseases, resistance to mechanical injury, resistance to elevated carbon dioxide/ low temperature and sugar contents. The correct choice for variety is very important for potato, carrot and onion. For instance, with carrot and swede, the variety that gives the most juicy grated product cannot be used in the production of grated products that should have a shelf-life of several days.

3. Cleaning /Washing/ Drying

Fruits or vegetables are cleaned and washed first before peeling, shredding or cutting to remove the adhering soil, dust or other undesirable materials stuck to them. It also removes the remains of any type of disinfectant sprayed on them while in the field. After washing the fruits or vegetables are wiped to remove the adhering moisture on their surface. After peeling/cutting, a second washing is done with ice cold water (<5^{0}C) that removes microbial growth tissue fluids and enzymatic oxidation. Preservative can be used in washing water for reduction of microbial load and reduce enzymatic activity. 100-200mg of chlorine or citric acid per lire of water is highly effective in washing water before and after peeling of cutting of fruits or vegetables. The effectiveness of chlorine can be enhanced by using a combination of low pH, high temperature, pure water and correct contact time. Alternatives to chlorine include chlorine dioxide, per acetic acid ozone, trisodium phosphate and hydrogen peroxide. Excessive free water should also be drained out by any suitable method such as centrifugation.

4. Peeling/Cutting/Shredding

Some fruits or vegetables, such as jackfruit, oranges, mangoes, potatoes, carrot need peeling and some don't need like spinach, sponge gourd etc. There are various methods of peeling such as hand peeling, steam peeling, lye peeling and industrial using carborundum drums. However, in the current practice of minimal processing , the peeling should always be as gentle a possible. The ideal method would be hand –peeling with a sharp knife. Carborundum peeled potatoes must be treated with a browning inhibitor, whereas water washing is adequate for hand peeled potatoes. Except hand peeling, all the methods such as carborundum, steam and lye peeling damage the cell walls of vegetables which accelerate the microbial growth and undesirable enzymatic changes. Hence, the cutting and shredding must be performed with highly sharp knives of blades and that should be made of food grade stainless steel. It has been foud that slicing with blunt blade of knife impairs quality retention because of the increased rupture of cells and release of tissue fluids. Mats and blades used in slicing should also be disinfected with 1% hypochlorite solution.

5. Antibrowning

One of the major problems in minimal processing is enzymatic browning in sliced or cut fruits or vegetables. The enzyme, PPO (polyphenoloxidase) is the main character that is responsible for this browning effect. To prevent enzymatic browning, washing with water and use of sulphites have long been used. However, the use of sulphites has some disadvantages such as side effects for asthmatics. Besides, some of the countries have also partly restricted the use of sulphites in food products. The most suitable alternative to sulphite is ascorbic acid. It is anoxidant and highly effective inhibitor of enzymatic browning. But, it works in the best way when applied with other substances like citric acid. Citric acid acts as a chelating agent and acidulant, both of which characteristics inhibit PPO. Since there is no one substitute for sulphites in preventing browning, alternatives are usually ascorbic acid-based combination. A typical combination may include a chelating agent (e.g. ascorbic acid), an acidulant (e.g. citric acid) and a chelating agent (e.g. EDTA).

6. Packaging

Packaging is one of the important operations in minimal processing of fruits and vegetables. The most studied and used packaging method for minimally processed fruits and vegetables is modified atmosphere packaging (MAP). The basic principle in MAP is that a modified atmosphere can be created passively by using properly permable packaging materials, or actively by using a specified gas mixture together with permeable packaging materials. The objective of the method is to create an optimal gas balance inside the package, where the respiration activity of a product could be reduced without having any detrimental effect on the stored produce. Generally, the aim is to have a gas composition where 2-5% CO_2, 2-5% O_2 and rest nitrogen. It also reduces water loss from the cut surfaces. Although MAP maintains visual quality by retarding browning, off-odours increase and lettuce crispness decreased during storage of salad products. With, recent packaging technology, it is possible to have product of good visual quality even at temperature-abuse conditions. Packaging technology is indispensable for most fresh cut fruits and vegetables. The selection of the plastic film packaging material involves a balance between rate of respiration and gas permeability of the film. Many types of films are commercially available for fresh cut fruits or vegetables such as polyethylene (PE), polypropylene (PP), blends of PE and ethylene vinyl acetate and co-extruded polymers or laminates of several plastics.

References

Abe, K., and Watada, A.E. (1991). Ethylene absorbent to maintain quality of lightly processed fruits and vegetables. J. Food Sci., 56: 1589–1592.

Ahvenainen, R. and Hurme, E. (1994). Minimal processing of vegetables, In:

Brecht, J. (1995). Physiology of lightly processed fruits and vegetables. HortScience,30(1): 18–22.

Fellows, P.J. (2000). Food Processing Technology, Principles and Practice. CRC Press Woodhead Publishing Limited. 236

Ke, D., and Saltveit, M.E. (1989). Carbon dioxide-induced brown stain development as related to phenolic metabolism in iceberg lettuce. J. Am. Soc. Hort., 114: 789–794.

Manvell, C.(1996). Opportunities and problems of minimal processing and minimally processed food". Paper presented to the EFFost Conference on minimal Processing of Food, Cologne, 6-9 November.

Ohlsson, T. and Bengtsson, N. (2000). Minimal Processing in the food industry. CRC Press, Woodhead Publishing Limited, England. 4

Perera, C.O., and Baldwin, E.A. (2001). Biochemistry of fruits and its implications on processing. In: Fruit Processing: Nutrition, Products & Quality Management, 2nd ed. Arthey, D., and Ashurst, P.R.,Eds. Aspen Publisher, Inc., Garthersburg, MD

Rahman, M.Shaifur. 2007. Handbook of Food Preservation, Second Edition. Taylor & Francis Group NW . 137

Sarawathy, S, Preethi, T.L., Balasubramanyan, Suresh ,J., Revathy, N. and Natrajan, S. (2008). Post Harvest Management of Horticultural Crops. AGRIBIOS(INDIA), Jodhpur. 473.

Watada, A.E., Abe, K., and Yamauchi, N. (1990). Physiological activities of partially processed fruits and vegetables. Food Technol., 44(5):116–122.

Wills, R.B.H., McGlasson, W.B., Graham, D. and Joyce, D.C. (1998). Post Harvest: An introduction to the physiology and handling of fruit, vegetable and ornamentals. UNSW Press. 38.

7

Radiation Processing of Food

Vasudeva K.R.* and Manjunath Shetty[1]

[*]Professor, [1]PhD Scholar, Dept. of Postharvest Technology, College of Horticulture UHS Campus, GKVK (P), Bengaluru-65

Need to Preserve Food

Food is vital for human existence. Conservation and preservation of food is a prerequisite for food security and it provides economic stability and self-reliance to a nation. The need to preserve food has been felt by mankind since time immemorial. The seasonal nature of production, long distances between production and consumption centers and rising gap between demand and supply have made this need even more relevant today. The hot and humid climate of a country like India is quite favorable for the growth of numerous insects and microorganisms that destroy stored crops and cause spoilage of food every year. Spoilage can also occur due to chemical and physiological changes in stored foods. Sea-foods, meat and poultry may carry harmful microbes and parasitic organisms that cause illnesses associated with their consumption. As in other parts of the world, India has also practiced various methods of food preservation such as sun drying, pickling and fermentation which were supplemented with more energy consuming techniques such as refrigeration, freezing and canning. Each of these methods has its merits and limitations. Man has always been on the search for newer methods to preserve foods with least change in sensory qualities. Radiation processing of food is one of the latest methods developed for this purpose.

Radiation processing of food involves exposure of food to short wave radiation energy to achieve a specific purpose such as extension of shelf-life, insect disinfestations and elimination of food borne pathogens and parasites. In comparison with heat or chemical treatment, irradiation is considered a more effective and appropriate technology to destroy food borne pathogens. It offers a number of advantages to producers, processors, retailers and consumers.

What is Food Irradiation?

Food irradiation is a promising new food safety technology that can eliminate disease-causing microorganisms such as *E. coli* O157:H7, Campylobacter, and Salmonella from foods.

The Food and Drug Administration has approved irradiation of meat and poultry and allows its use for a variety of other foods, including fresh fruits and vegetables, and spices. The agency determined that the process is safe and effective in decreasing or eliminating harmful bacteria. Irradiation also reduces spoilage bacteria, insects and parasites, and in certain fruits and vegetables it inhibits sprouting and delays ripening. The effects of irradiation on the food and on animals and people eating irradiated food have been studied extensively for more than 40 years. These studies show clearly that when irradiation is used as approved on foods:

- Disease-causing microorganisms are reduced or eliminated
- The nutritional value is essentially unchanged
- The food does not become radioactive

Irradiation is a safe and effective technology that can prevent many food-borne diseases.

The Food Irradiation Process

The radiation used in processing materials is limited to radiation from high-energy gamma rays, X-rays and accelerated electrons. These types of radiation are called "ionizing" because their energy is high enough to dislodge electrons from atoms and molecules and to convert them to electrically charged particles called ions.

- **Accelerated electrons:** The electron beam is a stream of high energy electrons, propelled out of an electron gun. This electron gun apparatus is a larger version of a standard television tube. The electron beam generator can be simply switched on or off. There are no radioactive materials in the process. The electrons can penetrate food only to a depth of three centimeters, or a little over an inch, so the food to be treated must be no thicker than that to be treated all the way through. Two opposing beams can treat food that is twice as thick. E-beam medical sterilizers have been in use for at least fifteen years.
- **Gamma rays and X-rays** form part of the electromagnetic spectrum, like radio waves, microwaves, ultraviolet and visible light rays. Gamma rays and X-rays are in the short wave length, high energy region of the spectrum. Both Gamma and X-rays can penetrate foods to a depth of several feet.

1. **X-rays** are caused by atomic transitions and they are usually less energetic than gamma rays. They have the same properties and effects on materials, their origin being the main difference between them. X-rays with varying energies are generated by machines. The X-ray machine is a more powerful version of the machines used in many hospitals and dental offices to take X-ray pictures. To produce the X-rays, a beam of electrons is directed at a thin plate of gold or other metal, producing a stream of X-rays. Like gamma rays, X-rays can pass through thick foods, and require heavy shielding for safety. However, like E-beams, the machine can be switched on and off, and no radioactive substances are involved. Four commercial X-ray irradiation units have been built throughout the world since 1996.
2. **Gamma rays** with specific energies normally come from the spontaneous disintegration of radionuclides. Naturally occurring and man-made radionuclide's, also called radioactive isotopes or radioisotopes, are unstable, and emit radiation as they spontaneously disintegrate, or decay, to a stable state. The radionuclide used almost always for the irradiation of food by gamma rays is cobalt- 60. It is produced by neutron bombardment in a nuclear reactor of the metal cobalt-59, then doubly encapsulated in stainless steel "pencils" to prevent any leakage during its use in a radiation plant. Cobalt-60 has a half-life of 5.3 years. This technology has been used routinely for more than thirty years to sterilize medical, dental and household products, and it is also used for radiation treatment of cancer. Radioactive substances emit gamma rays all the time. Whennot in use, the gamma ray "source" is stored in a pool of water which absorbs the radiation harmlessly and completely. To irradiate food or some other product, the source is pulled out of the water into a chamber with massive concrete walls that keep any rays from escaping. Medical products or foods to be irradiated are brought into the chamber, and are exposed to the rays for a defined period of time. After it is used, the source is returned to the water tank. Only certain radiation sources can be used in food irradiation. These are the radionuclide's cobalt-60 or cesium-137 (used very rarely); X-ray machines having a maximum energy of five million electron volts (MeV); or electron machines having a maximum energy of 10MeV. Energies from these radiation sources are too low to induce radioactivity in any material, including food.

Radiation dose is the quantity of radiation energy absorbed by the food as it passes through the radiation field during processing. It is measured in Gray (Gy) [one Gray equal one Joule of energy absorbed per kilogram of food being irradiated] or in rad (1 Gy = 100 rads). International health and safety authorities have endorsed the safety of irradiation for all foods up to adose level of 10,000 Gy (10 kGy).

Application of Radiation Processing of Food

Interest in the practical application of the process is emerging for many reasons. High food losses caused by insect infestation, microbial contamination and spoilage; mounting concern over food borne diseases, harmful residues of chemical fumigants and the impact of these chemicals on the environment, the stiff standards of quality and quarantine restrictions in international trade are some of the reasons. Though irradiation alone cannot solve all the problems of food preservation, it can play an important role in reducing post-harvest losses and use of chemical fumigants. On the basis of radiation dose, applications of radiation can be classified into:

Low Dose Applicat. Low Dose Applications	
Sprout inhibition in bulbs and tubers	0.03 - 0.15 kGy
Delay of fruit ripening	0.25 - 0.75 kGy
Insect disinfestation including quarantine treatment and elimination of food borne parasites	0.25 - 1 kGy

- **Medium Dose Applications**

Reduction of spoilage microbes to improve shelf-life of meat, poultry and seafoods under refrigeration	1.5 - 3 kGy
Elimination of pathogenic microbes in fresh and frozen meat, poultry and seafoods	3 - 7 kGy
Reducing number of microorganisms in spices to improve hygienic quality	10 kGy

• High Dose Applications

Sterilization of packaged meat, poultry and their products which are shelf-stable without refrigeration	25 - 70 kGy
Sterilization of hospital diets	25 - 70 kGy

Types of irradiation

a) Radappertization (30-40 kGy): Equivalent to sterilization or "commercial sterility," as it is understood in the canning industry.

b) Radicidation (2.5-10 kGy): Equivalent to pasteurization of milk. (Reduction in the number of viable non-spore forming pathogens).

c) Radurization (0.75-2.5 kGy): Considered as equivalent to pasteurization. Enhancement of the food quality by substantial reduction in the numbers of viable spoilage microbes.

The Benefits and Limitations of Radiation Processing of Food

Benefits	Limitations
Radiation processing is a cold process and therefore, unlike heat, it can be used on agricultural commodities without changing their fresh-like character	Radiation processing is a need-based technology and cannot be applied to all kinds of foods
Radiation processing does not alter significantly nutritional value, flavour, texture and appearance of food	Radiation processing cannot make a bad or spoiled food look goo
Radiation using Cobalt-60 cannot induce any radioactivity in food and does not leave any harmful or toxic radioactive residues on foods as is the case with chemical fumigants	It cannot destroy already present pesticides and toxins in foods
Due to the highly penetrating nature of the radiation energy, it is a very effective method	Amenability of a particular food commodity to radiation processing has to be tested in a laboratory
Prepackaged foods can be treated for hygienization and improving shelf-life	Only those foods for which specific benefits are achieved by applying appropriate doses, and those duly permitted under the Prevention of Food Adulteration Act (PFA) Rules, 1955, can be processed by radiation.
The radiation processing facilities are environment friendly and are safe to workers and public around	

Control and Regulations

Control of food processing in all types of facilities involves the use of accepted methods of measuring the absorbed radiation dose, the distribution of that dose in the product package, and of monitoring of the physical parameters of the process. Dosimetry for food irradiation is analogous to temperature monitoring in thermal processing, with dose meters being the analogs of thermocouples in that they are employed to provide an accurate measure of the rate of energy delivery, total energy delivered or absorbed and the over dose ratio. There are several types of dose meters, the most common ones being based on a chemical change that is linear within a practical dose range. Dosimetry is the keystone of the proper radiation processing. Careful dosimetry is required to ensure that a technologically useful dose has been applied, while maintaining the best possible dose uniformity ratio. Therefore, prior to commissioning of the plant, extensive dosimetric calibrations of the irradiator are carried out and followed during the processing by routine dosimetry. Because it is not possible to distinguish irradiated from non-irradiated products by sight, smell or taste, it is important that appropriate indicator devices which undergo radiation induced colour change be attached to each container/package, and that physical.

Economics of Radiation Processing of Foods

The design of the facility and therefore the economics depend on the parameters associated with radiation processing of food. These parameters include dose range, nature of food product, density of product, irradiator efficiency, annual operating hours, throughput, Cobalt-60 loaded, replenishment of Cobalt-60 required and the number of handlers employed, besides capital cost and cost of operation, maintenance, etc. In general, a radiation processing facility may cost between 5-10 crores. The cost of processing charged to product is expected to be less than 5% of the cost of the commodity. Past experience has shown enough market demand to sustain this activity.

International Status

The Joint Expert Committee of the Food and Agricultural Organization (FAO), World Health Organization (WHO), and International Atomic Energy Agency (IAEA), in 1980 concluded that food processed by radiation up to a dose of 10 kGy is safe for human consumption and does not pose any special nutritional, microbiological or toxicological problems. These conclusions have also been reviewed and endorsed later by the similar Joint Expert groups set up from time to time and also by a number of scientific and technical associations and bodies Today over 30 countries worldwide, including India, have approved the use of radiation processing of foods for over 100 food items and some of these countries are applying the technology on a limited commercial scale.

References

Akram, K. and Kwon, J. H., 2010. Food irradiation for mushrooms: A review. Journal of the Korean Society for Applied Biological Chemistry, 53(3): 257-265.

Arvanitoyannis, I.S., Stratakos, A.C. and Tsarouhas, P., 2009. Irradiation applications in vegetables and fruits: A review. Critical reviews in food science and nutrition, 49(5): 427-462.

Loaharanu, P. and Thomas, P., 2020. Irradiation for food safety and quality.CRC Press.

Molins, R. A, 2001. Food Irradiation: principles and applications. John Wiley and Sons.

Thayer, D. W, 1990. Food Irradiation: benefits and concerns. Journal of food quality, 13(3): 147-169.

8

Nutraceutical and Functional Foods

Rakesh Sharma*[*] and Dev Raj*[**]

**Department of Food Science and Technology, Dr YS Parmar University of Horticulture, Nauni, Solan, Himachal Pradesh-173 230, India*

***Department of Post Harvest Technology, ASPEE College of Horticulture Navsari Agricultural University, Navsari, Gujarat-396 450, India*

Introduction

The health and wellness of human being is largely dictated by the consumption of nutritious foods. Understanding the role of foods in maintaining health and cure of diseases has its roots in the ancient Hippocratic and Vedic texts as well as in the canons of Traditional Chinese Medicine (Sharma *et al.,* 2012). Presently, the transition from a traditional to modern lifestyle, consumption of diets rich in fat and calories combined with a high level of mental stress has been the major cause of lifestyle related diseases like coronary heart disease, obesity, diabetes, cancer and hypertension (Moczara *et al.,* 2012; Sharma *et al.*, 2018). Various studies have linked foods as helpful in combating a number of such degenerative diseases (Sharma, 2015).Therefore, the tenet "*Let food be thy medicine and medicine be thy food*", quoted by Hippocrates about 2,500 years ago is receiving overwhelming interest. There are number of scientific reports emphasizing the importance of consuming biologically active ingredients present in fruits, vegetables, whole grains and legumes to reduce the risk of chronic diseases (Hasler 2002; Das *et al.,* 2011; Sharma and Sharma 2016). Such biologically active ingredients are now popularly known as neutraceuticals. Whereas, the functional foods are the foods that resemble traditional foods but possess demonstrated physiological benefits. In some countries, however, functional foods and nutraceuticals are used interchangeably. Regardless, the main focus of such products is to improve health and reduce disease risk through prevention. The main difference of this category with pharmaceuticals is that they are multi-targeted mixtures and present at low concentration while pharmaceuticals are uni-targeted pure compounds with high dose use. Hence, an overview on the concept of nutraceutical and functional foods, their classification, development and regulations is summarized in this article.

Concept of Nutraceutical and Functional Foods

Various terms have been used though interchangeably to designate foods for disease prevention and health promotion. The concept of "nutraceutical" arose first in the survey from U.K., Germany and France, where a diet was rated higher by the consumers, than exercise or hereditary factors to achieve a good health (Pandey *et al.,* 2010). The term "nutraceutical" was coined from "*nutrition*" and "*pharmaceutical*" by Stephen De Felice, founder and chairman of the Foundation for Innovation in Medicine (FIM), Cranford in 1989 and referred it to as "*any substance that is a food or a part of a food which provides medical or health benefits including the prevention and treatment of disease*" (De Felice, 1995).*Nutraceutical is a product isolated or purified from foods that is generally sold in medicinal forms not usually associated with foods.* There is an extremely wide range of functional ingredients or nutraceutical or bioactive compounds which are knowingly or unknowingly being consumed via one or the other food that are associated with beneficial effects on human health (Hasler, 2002). Plant based nutraceutical compounds are known as phytochemicals or phytoceuticals.

According to the Food, Drug and Cosmetic Act (1968) food is "*an article that provides taste, aroma or nutritive value*". A food may be considered to be *functional* if it contains a food compound/ component (whether a nutrient or not) which affects one or more identified functions in the body in a positive manner. The concept of *functional food* was first developed in Japan in the mid-1980s when faced with escalating health care costs and certain foods with documented health benefits were approved in hope of improving the health of the nation's aging population. These foods are now known as *"food for specific health use"* (FOSU). Since then, various terms like designer food, dietary supplements, medical food, organic food, health food etc. have been used though interchangeably to designate foods for disease prevention and health promotion(Arai, 1996). However, there is no generally accepted definition of functional foods but according to Health Canada "*a functional food is similar in appearance to, or may be, a conventional food that is consumed as a part of a usual diet, and is demonstrated to have physiological benefits and/or reduce the risk of chronic disease beyond basic nutritional functions*". Whereas, the US Academy of Science has defined functional foods as "*foods that encompass potentially healthful products including any modified food or food ingredient that may provide a health benefit beyond the traditional nutrients it contains*". According to the American Dietetic Association (ADA),the functional foods comprised of (i) conventional foods containing naturally occurring bioactive substances (eg. Dietary fiber), (ii) foods enriched with bioactive substances (eg. probiotics, antioxidants) and (iii) synthesized food ingredients introduced to traditional foods (eg. prebiotics). The definitions as given by some scientific and regulatory bodies globally are given in Table 1.

Table 1. Definition of functional foods as given by various scientific and regulatory bodies

Institute of Food Technologists (IFT), USA	Foods and food components that provide a health benefit beyond basic nutrition (for intended population) including conventional foods, fortified, enriched or enriched foods and dietary supplements. They provide essential nutrients often beyond quantities necessary for normal maintenance, growth and development and/or other biologically active components that impart health benefits or desirable physiological effects.
International Food Information Council (IFIC)	Any food or food component that may provide a health benefit, beyond basic nutrition.
Food and Nutrition Board of the National Academy of Science (USA)	Potentially healthful products that may include any modified food or food ingredient that may provide a health benefits beyond the traditional nutrients it contains.

The concept of functional food is not new in India and it is more closely associated with Indian System of Medicine (ISM) popularly referred as *Ayurveda* in which many plant and milk based products are used for curing the various diseases since ages. Some of the biologically active ingredients/ nutraceutical compounds identified are listed in Table 2.

Table 2. Important bioactive/ functional/ nutraceutical compounds along with their source and potential health benefits

Class / components	**Source (s)**	**Potential health benefits**
1. Fatty acids		
Conj. Linoleic Acid (CLA) and n-3 FA(DHA, EPA)	Cheese, Milk & Meat products; Fish oils, mustard, linseed, rapeseed, tree nuts	Improve body composition, reduce cancers, reduce CVD, improve mental & visual function
2. Polyphenols		
Anthocyanidine, Catechins, Flavones & Flavonone	Fruits (Blue and black berries), Tea, mustard cake, rape seed, vegetables, soybean	Neutralizes free radicals, antioxidant and reduce risk of cancer
3. Probiotics / Prebiotics / Synbiotics		
Lactobacillus, Bifidobacteria, Fructo - oligosaccharides	Yogurt, Dahi, Probiotic drink, Whole grains, onions, Pro & Pre-biotics	These are dietary supplements which improve the GI health

4. Phytoestrogen		
Isoflavones- Daidzein &Genistein, Lignans	Soybean, flax, lentilseed, maize, vegetables	Reduce menopause symptoms, improve bone &brain health, boost immune system, Reduce cancer and heart diseases
5. Caroteinoids		
β- carotene, Luteine, Zeaxanthine, Lycopene	Carrots, vegetables, mango, papaya, marigold, eggs, citrus, tomato & tomato products etc.	Neutralizes free radicals, improve Healthy vision, reduce prostate cancer
6.Dietary fiber		
Insoluble fiber, whole grain, β-glucan	Wheat bran, cereal grains, oats	Reduce breast cancer, reduce CVD, healthy digestive system
7. Saponins	Soybeans, chickpea	Lower cholesterol, anti- cancer
8. Soy protein	Soybean and soy products	Reduce risk of CHD
9. Isothiocyanates (Sulforaphane)	Cauliflower, cabbage, broccoli etc.	Detoxification of undesirable compounds and improve antioxidant defense system

Source: De Felice (1995); Hasler (2002); Sharma *et al.* (2012); Sharma and Sharma (2016)

Functional Food *vs* Nutraceutical

Functional foods are the food products that resemble traditional foods but possess demonstrated physiological benefits. However, nutraceuticals are commodities derived from foods, but are used in the medicinal form of pills, capsules or liquids and again render demonstrated physiological benefits. The growing scientific evidences linking role of diet in maintaining good health and preventing diseases; aging population; adverse effects of drugs or pharmaceuticals in long run and the rising health care costs and/or medical expenditures are some of the factors for shift of a large population throughout the world towards functional foods and nutraceutical. To qualify as functional food, the food should meet following conditions:

- Food (not capsule, tablet or powder) should have been derived from naturally occurring ingredients.
- It can and should be consumed as part of daily diet.
- When ingested, it should have the capacity to regulate or influence certain physiological processes, such as improvement of biological defense mechanisms i.e. immune enhancer.

The critical differences between a functional food and nutraceutical have been illustrated below for easy understanding of these two terms.

Functional Food	Nutraceutical
• Functional foods should look as traditional/ conventional food products e.g. Fruit juice, flour etc.	• Nutraceutical generally available as extracts, powders, tablets, capsules, concentrates etc. e.g. carotene, Vit C etc.
• Should contain nutraceutical (bioactive compound) as one of their constituents	• These are constituents of the food or part of food, which when incorporated to basic food make it functional
• Should be a part of our basic food stuff and daily diet for maintaining good health	• Necessarily should not be a part of our daily meal and should not be consumed in excess

Categories of Nutraceutical

Nutraceutical can be classified based on the origin or source and nature of the beneficial bioactive ingredient/ compound and/or their therapeutic benefits claimed. Major categories of nutraceutical are illustrated in Fig 1.

- Nutrients: Substances which have established nutritional functions e.g. Vitamins, Minerals, Amino Acids, Fatty acids etc.
- Herbal/ Phytochemicals: Herbs or botanical products in the form of extracts or concentrates. eg. Aloe vera
- Dietary Supplements: Those contain "dietary ingredients" and serving specific functions such as meal replacement, weight loss. e.g. Metabolites, probiotics, prebiotics, antioxidants, enzymes, etc.
- Traditional nutraceutical: These are the foods which are natural whole foods without having any change to their composition. Many fruits, vegetables, grains, fish, dairy and meat products contain several natural components that deliver benefits beyond basic nutrition such as lycopene in tomatoes, omega-3-fatty acid in Salmon, golden rice etc.

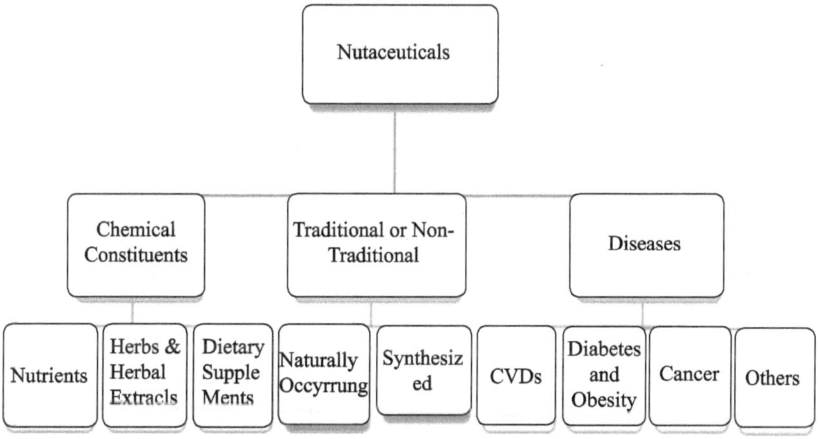

Fig. 1. Major categories of nutraceutical compounds

- Non-traditional or synthesized neutraceuticals: These are the foods resulting from various manufacturing techniques such as β-carotene enriched rice, vitamins enriched broccoli and soya bean etc.
- Potential and established nutraceutical: More broadly, nutraceuticals can also be grouped in two categories (Pandey *et al.*, 2010) as potential and established nutraceutical. A potential nutraceutical could become an established one only after efficient clinical data of its health and medical benefits are obtained e.g. Nutraceuticals for CVDs.
- Nutraceutical for disease management: Nutraceuticals which are associated with prevention or cure of diseases. Some of the neutraceuticals having established role in prevention of diseases are summarized in Table 3.

Table 3. Role of various nutracuticals in prevention or cure of diseases

Diseases	**Nutraceuticals**
Cardiovascular	Antioxidants, Dietary fibers, Vitamins, Omega-3-PUFA together with physical exercise
Obesity	Linoleic acid (CLA), Capsaicin, *Momordica Charantia* (MC), *Psyllium* fibers & other dietary supplements with added Glucomannan, Chitosan etc.
Diabetes	Omega-3-PUFA, Dietary fibers, Bitter gourd extract, Antioxidant (LA), Phytoestrogens (Isoflavones) etc.
Cancer	Lutein- rich food, Antioxidants, (Flavonoids, Lycopene, Beta-carotenoid), Phenolics (Curcumin), Thio-sulfonates, Fibers, Phytoestrogens etc.
Inflammation and allergy	Glucosamine, Quercetin (found in Onions, red wine and green tea) reduce the inflammation and allergy symtoms
Vision improving agents	Lutein, Zeaxanthin
Osteoarthritis	Glucosamine (GLN) and chondroitin sulfate (CS) alleviate symptoms of Osteoarthritis

Source: Gupta *et al.* (2010), Sharma and Joshi (2015), Sharma and Sharma (2016)

Categories of Functional Foods

The functional foods may comprise ofconventional foods containing naturally occurring bioactive substances, foods enriched with bioactive substances and synthesized food ingredients introduced to traditional foods (e.g. prebiotics). Broadly, functional foods can be classified into two main classes as follows:

- Basic / conventional foods: These are the foods that naturally contain biologically active, non-nutrient compounds that provide health benefits. These compounds are mostly phytochemicals.

- Formulated foods: This category includes the food products specially formulated to have higher amounts of the bioactives/ nutraceuticals than that would naturally occur in a product. These are mostly processed products specially designed to meet needs of a specific target group of population and are also called as modified food, designed foods, medical food and foods for special dietary uses. The main categories of functional foods along with their examples are given in Table 4.

Table 4. Categories of functional foods and some of their examples

Functional food category	Selected functional food examples
Conventional foods (whole foods)	Tomatoes, Carrot, Mango, Grape, Garlic, Nuts, Kale, Raspberries, Broccoli etc.
Modified foods: Fortified, Enriched and Enhanced	Calcium-fortified orange juice, Fe fortified flour,Folate-enriched breads, high fibre bread,Enriched snack bars, yogurts, teas, bottled mineral water, fish oils, low sodium foods etc.
Medical foods	Phenylketonuria (PKU) formulas free of phenylalanine
Foods for special dietary use	Infant foods, gluten-free and lactose-free foods, weight-loss foods, weight- gain foods, immune enhancer, probiotics etc.

Extraction of Bioactive/ Nutraceutical Compounds

The extraction techniques for these biologically active compounds or nutraceutical have been widely investigated to obtain bioactive compounds from these plants species. Among various techniques, traditional methods such as Soxhlet extraction which has been used for many decades is very time consuming and require relatively range quantities of solvents but are still used on commercial scale for extraction of various bioactive compounds. However, recently, following new techniques viz. enzyme-assisted extraction, ultra sound assisted extraction, microwave assisted extraction, accelerated solvent extraction and super critical fluid extractionhave come up which are now preferred because of the low organic solvent requirement, less time consumption and lowered pollution (Wang and Weller, 2006).

Development of Functional Foods

The development of functional foods involves the application of various food processing techniques and approaches, depending on the objective and type of functional food to be manufactured. Some of the techniques/ approaches are outlined below:

- Fortifying /enrichment of food with bioactive compounds
- Low calories food products

- Combining food ingredients rich in nutraceutical
- Fermentation of food products
- Prebiotic, probiotic and symbiotics
- Microencapsulation of functional ingredients
- Nanotechnology

The commercial success of the functional foods will depend upon various factors such as selection of ingredient, level of addition, ingredient interaction with other food constituent and its impact on flavor and texture, retention of beneficial activity etc. (Sharma 2015). The various steps considered in developing a functional food product are illustrated in Fig.

Step 1: Identify relationship between food component and health benefit

- Identify the bioactive compount (s) to be incorporated in a basic food stuff

↓

Step 2: Demonstrate efficacy and determine intake level necessary to achieve desired effect

- Assess stability and bioavailability of compound (s) through analytical methods
- Demonstrate efficacy using biological tools and biomarkers
- Determine intake level of functional food for desired effect

↓

Step 3: Demonstrate safety at efficacious levels

- Assess safety if component is new to food use
- Consider food additive used for safety issues
- Address potential allergenicity, if necessary

↓

Step 4: Develop or select suitable food vehicles for bioactive component

↓

Step 5: Demonstrate scientific sufficiency of evidence for efficacy

- Submit evidence to FDA for claim approval, after having the scientific evidences, if necessary

↓

Step 6: Communicate benefits to consumers through different means

↓

Step 7: Conduct in market confirmation of efficacy and safety

- Monitor efficacy vis-a-vis intake level of the fnctional food
- Monitor safety issues and report/address if any noticed

Fig. 2. Process for bringing the functional foods to market

Labeling, Health Claims and Regulations

Three categories of health claims have been identified viz. therapeutic, risk reduction and structure/function.

- **Enhanced function/ Therapeutic claims** that a product can cure, treat, mitigate or prevent a disease or condition that fall into the therapeutic claim category.
- **Disease risk reduction claim** could be defined as significantly altering a major risk factor or factors recognized to be involved in the development of a chronic disease or abnormal physiological conditions through product use. These claims are also either product specific or generic.
- **Structure/ Function claims** are the claims about the biological roles of nutrients. Generally, such claims are permitted for general and recognized roles of nutrients in the maintenance of good health and normal growth and health and development. The US Dietary Supplement Health and Education Act (DSHEA) recognize it as the role of a nutrient or dietary ingredient intended to affect the structure or function in human body. Manufacturers may make statements about a food/ compounds' effect on the structure or function of the body e.g."calcium builds strong bones." But, unlike health claims, structure/function claims do not deal with disease-risk reduction and FDA does not pre-approve or authorize structure/function claims (Pandey *et al.*, 2010). In the year 2006, Food Safety and Standards Act was notified by the Government of India. In the Act, the category 'Food for Special Dietary Uses or Functional Foods or Nutraceuticals or Health Supplements' was defined under Section 22 of the Act. The provisions of this section recognizes formulations in the form of powders, capsules, tablets, liquids or jelly and other dosage form, except injectables. It is the great step in this direction, however the section makes it clear that such products can only make certain health benefits or promotion claims as may be permitted by the regulation made under the Act. Besides this, the other agencies which regulate such type of food products in India are Health Foods and Dietary Supplements Association (HADSA) through Dietary Supplement Health and Education Act, National Institute of Nutrition (NIN), Food and Drug Toxicology Research Centre (EDTRC) and National Nutrition onitoring Bureau (NNNB).

Classification of Functional Foods

The functional foods comprised of (i) conventional foods containing naturally occurring bioactive substances (eg. Dietary fiber) (ii) foods enriched with bioactive substances (eg. probiotics, antioxidants) and (iii) synthesized food ingredients introduced to traditional foods (eg. prebiotics). Functional foods can be classified into different categories depending upon the method of preparation.

1. Conventional foods (whole foods)
2. Probiotics (micro-organisms incorporated food)
3. Prebiotics (fibre incorporated foods)
4. Synbiotics (micro-organisms and fibre incorporated food)
5. Designer foods (Neutraceuticals incorporated food)

1. Conventional Foods

A natural food and especially an unprocessed one (vegetable or fruit, animal food, diary food, sea food etc.) is called whole food. When the broadest, least restrictive definition is employed, the use of functional foods for promoting health and relieving symptoms is as old as the practice of medicine specific dietary recommendations for treating or preventing various types of illness have been documented in Hippocratic and Vedic texts and the canons of traditional Chinese medicine. Plant foods such as fruits and vegetables contain many bioactive compounds which have the potential to delay the onset of many age-related diseases (Table 7 to 8). Many fruits and vegetables are rich source of vitamins like vitamin C, vitamin E and carotenoids, collectively known as 'antioxidant vitamins'. These vitamins are reported to exert their protective action by free-radical scavenging mechanisms for the prevention of oxidative reactions which otherwise lead to several degenerative diseases including cancer, cardiovascular diseases etc.. Carotenoids like lycopene, β-carotene, lutein, zeaxanthin are known to be the most efficient singlet oxygen quencher in the biological systems without the production of any oxidizing products. In 1992, a review of 200 epidemiological studies showed that cancer risk in people consuming diets high in fruits and vegetables was only one-half that in those consuming few of these foods. It is now clear that there are components in a plant-based diet other than traditional nutrients that can reduce cancer risk. Polyphenols, a large group of phytochemicals like flavonols, flavones, flavanones and anthocyanins, are produced by plants as secondary metabolites to protect them from photosynthetic stress, reactive oxygen species. The most commonly occurring polyphenols in food include flavonoids and phenolic acids. These are of current interest because they are reported to possess antioxidant, anti-inflammatory, anti-microbial, cardioprotective activities and play an important role in the prevention of degenerative diseases (Table 9).

Table 7. Potential benefits of important functional components of foods

Component	Food source	Potential benefit
Carotenoids		
Alpha-carotene	Carrots	Neutralize free radicals, which may cause Damage to cells
Beta-carotene	Fruits, vegetables	-Do-
Lutein	Green vegetables	Reduce risk of macular degeneration
Lycopene	Tomato products (ketchup, sauces)	Reduce risk of prostate cancer
Dietary fibre		
Insoluble fibre	Wheat bran	Reduce risk of breast or colon cancer
Beta-glucan	Oats	Reduce risk of cardiovascular disease
Soluble fibre	Psyllium	Reduce risk of cardiovascular disease
Fatty acids		
Omega-3 fatty acids-dha/epa	Tuna and other fish oils	Reduce risk of cardiovascular disease, Improve mental, visual functions
Conjugated linoleic acid cla	Cheese, meat products	Improve body composition, Decrease risk of certain cancers
Flavonoids		
Anthocyanidins	Fruits	Neutralize free radicals; reduce risk of cancer
Catechins	Tea	-Do-
Flavanones	Citrus	-Do-
Flavones	Fruits/vegetables	-Do-
Plant sterols		
Stanol ester	Corn, soy, wheat, wood oils	Lower blood cholesterol levels by inhibiting cholesterol absorption
Phytoestrogens		
Isoflavones: • daidzein • genistein	Soybeans and soy-based foods	Help in menopause symptoms, such as hot flashes, protect against heart disease and some cancers; lower LDL and total cholesterol
Lignans	Flax, rye, vegetables	-Do-
Tannins (proanthocyanidins)	Cranberries, cranberry products, cocoa, Chocolate	Improve urinary tract health, Reduce risk of cardiovascular disease

Table 8. Potential Health Benefits of important functional components

Ingredient	Potential Health Benefits
Calcium	Prevention of Osteoporosis Promotes healthy bones and teeth Protective against hypertension
Folic Acid	Protective against neural tube defects
Vitamin E	Protective against Heart Disease Protective against Cancer
Iodine	Protective against Goitre
Immuno-peptides	Protective against Cancer
Casein Phosphopeptides	Promotion of Calcium Absorption
Lactoperoxidase	Prevention of dental caries
Lactoferrin	Prevention of dental caries Prevention of gastro-intestinal infection Leukaemia therapy
Modified Lactoglobulin	Prevents milk protein allergies
Lactic Acid Bacteria	Promote a healthy gut
Immunoglobulins	Prevention of dental caries Prevention of gastro-intestinal infection Lowering of Cholesterol

Table 9. Functional compounds present in fruit and vegetable associated with health promotion

Type of functional compound/ typical component	Functional properties	Sources
• Carotenoids (alpha- and beta-carotene, cryptoxanthin, lutein, lycopene, zeaxanthin)	Quench singlet and triplet oxygen	Red, orange and yellow fruits and vegetables
• Glucosinolates, Isothiocyanates (indole-3-carbinol sulphoraphane	Increase xenobiotic metabolism and alter estrogen metabolism	Cruciferous vegetables
• Inositol phosphates (inositol hexaphosphate	Stimulate natural killer cell function	Soy foods
• Isoflavones (genistein, daidzein)	Estrogen agonist and antagonist, induce apoptosis	Soy foods
• Phenolic acids (gallic, ellagic, ferulic, chlorogenic, coumaric)	Antioxidant, enhance xenobiotic metabolism	Various fruits and vegetables
• Phytoallexins (resveratrol)	Antioxidant, platelet inhibition	Red wine
• Polyphenols (flavonoids, chalcones, catechins, anthocyanins)	Antioxidant, enhance xenobiotic metabolism, nhibit numerous enzymes	Various fruits and vegetables, Red wine and tea
• Sulfides (diallyl sulfides)	Antimicrobial, antioxidant	Onions and garlic

Source: Block *et al.* (1992), Delaquis and Mazza (1998), Yang *et al.* (2001), Sharma *et al.* (2003), Joshi *et al.* (2006)

Fermentation is one of the oldest forms of food preservation technology in the world carried out to enhance taste, aroma, texture, shelf-life and nutritional value of foods. The health enhancing effects of fermented foods have a place in folk medicine. Numerous fermented foods which are consumed around the world have health benefits that exceed those of their parent foods and can be considered functional foods in a broader sense. Various nutritional and health effects of fermented products over their unfermented counterparts are summarized in Table 10. Vinegars are produced by acetic acid fermentation of fruit juices. Different microorganisms like *Kloeckera apiculata*, *Metschnikowia pulcherrima*, *Candida* spp., *Pichia, Acetobacter* and *Gluconobacter* play role in cider vinegar production. Raw apple cider vinegar has been claimed to treat a number of different human health conditions such as arthritis, high blood pressure, all types of skin problems, chronic fatigue, heart burn and has also anti-glycemic effect.

Table 10. Fermentation as a tool for nutritional improvement

- The inclusion of micro-organisms along with nutrients in many fermented products enhances their nutritive value
- Due to break down of more complex compounds (catabolic) simple compounds such as sugars are formed which are easily digestible.
- Micro-organisms causing fermentation being metabolic too, synthesis several complex vitamins and other growth factors e.g. rivoflavin, vitamin B_{12} and precursor of vitamin C
- The fermentation can also enhance the nutritional value of plant materials, involving enzymatic splitting of cellulose, hemicelluloses and related indigestible polymers into simpler digestible compounds
- Fermentation results in the destruction of anti-nutritional factors improving the nutritive value of fermented foods indirectly.

Source: Joshi *et al.* (1999)

Lactic acid bacteria (LAB) involved in the natural fermentation of fruits and vegetables since ancient times due to its preservative action. LAB responsible for the natural fermentation of fruits and vegetables are within the genera of *Lactobacillus, Strptococcus, Pediococcus* and *Leuconostoc*. The reduction of pH and removal of large amounts of carbohydrates by fermentation are the primary preserving action of these bacteria. The proliferation of lactobacilli in fermented vegetables enhances their digestibility and increases vitamin levels. These beneficial organisms produce numerous helpful enzymes as well as antibiotic and anti-carcinogenic substances. Vegetables as raw material for manufacture of lactic acid fermented vegetables are naturally low in fat, cholesterol and energy, but are source of micronutrients (Vit. A, C and E, folic acid, selenium), fiber and other phytochemicals such as carotenoids, floavonoids and other phenolic compounds. The cruciferous vegetables (cabbage, cauliflower, broccoli, kale etc.) have good cancer fighting ability because they contain phytochemicals like isothocyanates and indoles.

2. Probiotic Foods

Probiotics are live bacteria and yeasts that are called "the good" microorganisms because they benefit the body, specifically the digestive system. These are generally present in the gut of human intestine. Probiotics available in some foods and dietary supplements are similar to micro-organisms that exist naturally in gut. Probiotics are bacteria that help maintain the natural balance of organisms (microflora) in the intestines. The normal human digestive tract contains about 400 types of probiotic bacteria that reduce the growth of harmful bacteria and promote a healthy digestive system. The largest group of probiotic bacteria in the intestine is lactic acid bacteria, of which *Lactobacillus acidophilus*, found in yogurt with live cultures, is the best known. Yeast is also a probiotic substance. So, the foods which contain such probiotic micro-organisms are called probiotic foods.

Food / preparation (dietary supplement) containing live bacteria (lactobacilli) that are taken orally to restore beneficial bacteria to the body are called probiotic foods. The history of probiotics dates back as far as the first intake of fermented milks, over 2,000 years ago however, the scientific interest in this area boosted from the work of Metchinkoff in 1907. A probiotic can be defined as live microbial feed supplement, which when administered in adequate amounts beneficially affects the host animal by improving its intestinal microbial balance (Fuller, 1992). The probiotics generally include the following categories of bacteria: -

- *Lactobacilli such as L. acidophilus, L. casei, L. brevis, L. delbrueckii subsp. bulgaricus*
- *Gram+ve Cocci such as Lactococcus lactis, Streptococcus salivarius subsp. thermophilus*
- *Bifidobacteria such as B. bifidun, B. infantis, B. longum, B. thermophilum.*

Probiotics are available in various forms as powder form, liquid form, gel or paste or granule forms, capsule forms etc.. Specific probiotics are generally used to treat gastrointestinal (GI) conditions such as lactose intolerance, acute diarrhea and antibiotic-associated GI side effects. Probiotic agents possess the properties of non-pathogenic, non-toxic, resistance to gastric acid, adherence to gut epithelial tissues producing antibacterial substances. There are evidences that administration of probiotics decreases the risk of systemic conditions, such as allergy, asthma, cancer and several other infections of the ear, urinary tract. Probiotics are also available as dietary supplements. Foods that contain probiotics include some juices and soy drinks, fermented and unfermented milk, buttermilk, some soft cheeses, miso, tempeh, kefir, kim chi, sauerkraut, and many pickles. As said, probably the most well known food product that contains probiotics is yogurt

The lactic acid bacteria (LAB) involved in the natural fermentation of fruits and vegetables are capable of producing inhibitory substances other than organic acids, which are antagonistic towards other microorganisms. These include hydrogen peroxide, diacetyl, secondary reaction substances and bacteriocins. Some of the bacteria possess the genetic capability to synthesize the compounds referred to as bacteriocins. Bacteriocins are defined as low molecular weight proteinaceous compound produced by many strains of starter bacteria which are bacteriocidal to closely related strains and species. Being produced from food grade bacteria, their stability under conditions of processing and degradation by proteolytic enzymes of human gastrointestinal tract, without any recognizable problems have make them ideal bio-preservatives for food usage. Some of the lactic acid bacteria, their bacteriocin type and inhibitory spectrum in given in Table 11 and 12.

Table 11. Important Probiotics micro-organisns and their species

Lactobcaillus sp.	***Bifidobacterium sp.***	***Other Lactic Acid Bacteria***	***Other Species***
L. acidophiuls	*B. adolescentis*	*Enterococcus faecalis*	Bacillus cereus
L. casei	*B. animalis*	*E. faecium*	*Saccahromyces boulardiii*
L. cellobiosus	*B. longum*	*Streptococcus thermophiuls*	
L. fermentum	*B. brevi*		
L. lactis	*B. bifidum*		
L. helveticus	*B. infantis*		
L. reuteri	*B. lactis*		
L. brevis *L. plantarum*			
L. curvatus			

Table 12. Lactic acid bacteria, their bacteriocin type and inhibitory spectrum

Sr No	Lactic acid bacteria	Bacteriocin type	Inhibitory spectrum
1.	*Lactococcus lactis* subsp. *Lactis*	Nisin	Many Gram positive bacteria; *Bacillus, Clostridium, Listeria, Micrococcus* etc.
2.	*Lactococcus lactis* subsp. *Cremoris*	Diplococcin	Wide range of Gram positive bacteria
3.	*Bacillus subtilis*	Subtilin	It has inhibitory spectrum similar to Nisin
4.	*Pediococcus acidilactici*	Pediocin1 Pediocin AcH	Some LAB and *Listeria, Clostridium sp.*
5.	*Pediococcus pentosaceus*	Pediocin A	Some LAB and *Listeria, Staphyllococci* and *Clostridium bolulinum*
6.	*Lactobacillus reuterii*	Reuterin	Gram +ve and –ve bacteria, yeast and fungi
7.	*Lactobacillus acidophilus*	Lactococin/ Acidolin/ Lactacin	Narrow spectrum antibiotic
8.	*Lactobacillus plantarum*	Lactolin/ Plantaricin	Related strains of LAB; *L. brevis, L acidophyllus* etc.
9.	*Streptomyces natalensis*	Natamycin/ Pimaricin	Effective against bacteria, viruses and actinomyces
10.	*Bifidobacterium sp.*	Bifidin	*Bifidobacteria, E. coli* and related strains
11.	*Propionibacterium shermanii*	Microgard	Gram +ve bacteria, some yeast and moulds

Source: Sharma *et al.* (2003); Joshi *et al.* (2006)

3. Prebiotic Foods

These are chemical or environmental precursors of the origin of life and exist before the origin of life. E.g. a food / drink in which fibre have been added to give health benefits are called prebiotic foods. Prebiotics are an alternative for probiotics. They are non digestible or low digestible food ingredients that benefit the host organism by selectively stimulating the growth or activity of one or a limited number of probiotic bacteria in the colon. Prebiotics are not digested in the upper part of gastrointestinal tract and reach the colon, where these are metabolized by bacteria residing in the colon and imparting beneficial effect on the host. This role is played by fermentable carbohydrates, which are not digested or poorly digested in the small intestine and stimulate, preferentially the growth of *Lactobacillus, Bifidobacteria* and some Gram+ve bacteria belonging to the probiotic bacteria administered to humans. The other potential functional effects of prebiotics are on the bioavailability of minerals and on lipid metabolism. In prebiotic production, complex carbohydrates pass through the small intestine to the lower gut where they become available for some colonic bacteria but are not utilized by the majority of the bacteria present in the colon. The main end products of carbohydrate metabolism are short chained fatty acids, namely acetate, butyrate and propionate which are further used by the host organism as an energy source. The most common oligosaccharides are inulin and its hydrolysates and oligotructanes. The consumption of prebiotics generally promotes the growth of Lactobacillus and Bifidobacteria in the gut, thus helping in metabolism. Vegetables like chicory roots, banana, tomato, alliums are rich in fructo-oligosaccharides, whereas oligosaccharides such as raffinose and stachyose are found in beans and peas. The potential health benefits of the prebiotics include improved lactose tolerance, anti-tumor properties, neutralization of toxins, and stimulation of intestinal immune system, reduction of constipation, blood lipids and blood cholesterol levels. A daily intake of 5¬20 g of insulin and oligosachharides promote the growth of bifidobacteria. However, consumption of large amounts of such oligosaccharides has also been reported to cause diarrhea, abdominal distension and flatulence.

Prebiotic oligosaccharides can be produced in three different ways (i) by extraction from plant materials, (ii) by microbiological synthesis, enzymatic synthesis and (iii) enzymatic hydrolysis of polysaccharides. Oligosacchaides act as prebiotics and they are also called as bifidogenic factors, because used by bifidobacteria or other LAB's.

4. Synbiotic Foods

Synbiotic is the word coined for the combined administration of specific probiotics to provide definite health benefits by synergistic action. A synbiotic is a product containing probiotic (fibre, pectin etc) and prebiotic (microbes) in which the prebiotic compound specifically favors for probiotic compound. Thus, combined mixtures of probiotics and prebiotics are often used because of their synergic effects on food products. Such mixtures are therefore called as '*Synbiotics*'.

5. Designer Food

Designer foods, coined in 1989, is used to describe foods that naturally contain or are enriched with non-nutritive, biologically active chemical compounds of plants that are effective in reducing cancer risk. Thus, the food product formulated to contain / enhance levels of the phyto chemicals is called designer foods. These foods are supplemented with food ingredients that are naturally rich in disease preventing substances. These are sometimes also known as Nutraceutical / Nutriceutical. Nutraceutical is the name given to a range of food supplements which are designed to nutritionally support certain illness, improve metabolic processes and correct biochemical balance etc. *Nutraceuticals* was also introduced in 1989 by the US Foundation for Innovation in Medicine to refer to "any substance that is a food or a part of a food and provides medical or health benefits including the prevention and treatment of disease". Thus, Nutraceutical is a product isolated from foods but sold in powders, pills and other medicinal forms not generally associated with food and are demonstrated to have physiological benefits or provide protection against chronic diseases. Vitamin pills, mineral pills, herbal pills/supplements, oil gel caps, phytochemicals, and zoochemicals are the important examples of nutraceuticals. They are also referred to as 'bioactives', indicating that they are physiologically active and as 'phytochemicals', since they are essentially of plant origin. Functional foods have also been given such general names as 'designer foods', 'pharma foods', 'health foods' or 'therapeutic foods'.

The term 'designer foods' is subjected to change based on nutritional labelling regulations, new developments, consumer perceptions and expectations. Therefore, it can be said that designer food is an evolving concept meant for reducing or delaying the risk of diseases.

Types of designer foods

1. Designer Meat Foods (Low Fat Meat Products; Low-Sodium Meat Products; Nitrite Reduced Meat Products; Meat Products with Designer Food Ingredients).
2. Designer Cereals/ Flour (high molecular-weight glutenins)
3. Designer Milk (more protein, less lactose and absence of b-lactoglobulin, Milk with Modified Fat, Milk with Modified Protein, Milk with Modified Carbohydrate).
4. Designer Eggs (vegetarian eggs and eggs with modified fat content)
5. Designer Fruits (anti-oxidant and flavonoide rich)

Conclusion and Future Strategy

There are many functional foods and neutraceuticals that are becoming increasingly available in the marketplace. Further, increasing health consciousness has been one of the most important stimulating factors for the rapid global growth of the nutraceutical and functional food industry. Presently, the research and development is this field is at the peak. Many developed countries have a government policy to advocate *5-plus a day* fruits and vegetables in their diets so as to curtail the ill effects of *junk food.* But, but there is a challenge for the functional food producers because such products should address the issue of sensory acceptability which is not necessary for the nutraceutical or pharmaceutical products. For nutraceutical compounds, clinical studies are required to provide the basis for health claims and to produce an impact on the consumers as well as on the functional food companies. In addition, processing by-products from plant sources are particularly rich in a number of bioactives at much higher concentration than those present in the main products and these can be used as value-added ingredients for developing functional foods.

References

Arai S. 1996. Studies on functional foods in Japan-State of the art. Biosci. Biotech. Biochem., 60: 9-15.

Das L, Bhaumik E, Raychaudhuri U and Chakraborty R. 2011. Role of nutraceuticals in human health. J Food Sci Technol. DOI 10.1007/s13197-011-0269-4

De Felice SL. 1995. The nutraceutical revolution: its impact on food and industry R & D. Trends Food Sci. Technol. 6: 59-61.

Gupta S, Chauhan D, Mehla K, Sood P and Nair A. 2010. An overview of nutraceuticals: Current scenario. J Basic Clinical Pharmacy, 001(002): 55-62

Hasler CM. 2002. Functional foods: Benefits, Concerns and Challenges. The J. Nutr. 132: 3772–3781.

Moczara C, Borgulyab G, Kovacsc E. and Rurikc I. 2012. Could primary care dietary intervention combined with lifestyle changes be effective in the cardiovascular prevention? Acta Alimentaria, 41(2): 248–256.

Pandey M, Verma RK, Saraf SA. 2010. Nutraceuticals: new era of medicine and health. Asian J Pharm Clinical Res 3: 11-15.

Sharma P C and Sharma R. 2016. Fruit and vegetable based health foods. In: Souvenir of ICAR-Sponsored Summer School on "Engineering and Technology Innovations in Developing Health Foods" (RK Gupta, Sangita Bansal and Dhritiman Saha, Eds). ICAR-CIPHET, Ludhiana (8-28 June, 2016). pp. 66-79.

Sharma Rakesh and Joshi VK. 2015. Novel Technology for Development of Low Calorie Functional Beverages from Fruits and Vegetables Using Non-Nutritive Sweeteners, Chapter 21. Functional Foods and Nutraceuticals- Sources and their Developmental Techniques (CS Riar, DC Saxena, S Singh V Nanda & N Jindal, Eds). NIPA, New Delhi, pp: 233-243.

Sharma Rakesh, Joshi VK and Abrol GS. 2012. Fermented fruit and vegetable products as functional foods- an overview. Indian Food Packer, 66(4): 45-53

Sharma Rakesh, Sharma Ranjana and Thakur Abhishek, 2018. Development and evaluation of Vitamin C enriched low calorie Aloe vera- aonla blended functional squash using stevioside. Indian J Horticulture, 75(2): 289-294.

Sharma Rakesh, 2015. Development of functional foods and nutraceutical for nutritional security and human health management: An Overview. In: Souvenir of National Symposium on Modern Agro-technologies for Nutritional Security and Health. YSP University of Horticulture and Foretry, Nauni Solan (HP), pp. 96-102.

Wang L and Weller CL. 2006. Recent advances in extraction of nutraceuticals from plants. Trends in Food Sci Technol 17: 300-312.

9

Thermal and Non-thermal Methods of Processing

F.M. Sahu and Dev Raj

Centre of Excellence on Post Harvest Technology,Navsari Agricultural University Navsari, Gujarat-396450, India

Introduction

It is the process of retaining food over a period of time without being contaminated by pathogenic organisms or chemicals and without losing its colour, texture, flavour and nutritious value. It involves preventing the growth of bacteria, fungi and other microorganisms, as well as retarding the oxidation of fats which causes rancidity.

Objective

- To increase its shelf life, retaining original nutritional values, colour, texture, and favour.

The shelf-life of a product is the period of time during which the food retains its qualitative characteristics.In practice, there are three important methods used to control or kill the microorganism for preservation of fruits and vegetables:

1. **Physical methods:** Mostly temperature control methods like low or high temperatures (thermal) and non- thermal processing like irradiation.
2. **Chemical methods:** It included certain antimicrobial chemicals such as sorbate, sodium benzoates, potassium metabisulphite (KMS), salt, sugar, vinegar.
3. **Biological preservation:** microbial growth is encouraged so as to carry fermentation to preserve foods.

Under the Physical methods, thermal and non-thermal food preservation methods are included.

(A) Thermal food preservation/processing - Adding or removing heat to food to destroy spoilage and pathogenic microorganisms and enzymatic destruction as well as alter shelf life of food and maintenance of its quality.

The term "thermal" refers to processes involving heat. The term thermal technologies encompass the processes and equipment especially designed to heat a food with the purpose of processing it. This term is often used in opposition to non-thermal technologies, where food is processed but virtually no heat is involved. Heating food is an effective way of preserving it because the great majority of harmful pathogens are killed at temperatures close to the boiling point of water. In this respect, heating foods is a form of food preservation comparable to that of freezing but much superior to it in its effectiveness.

(B) Non-thermal food processing

(A)	Thermal processing methods	(B)	Non-thermal processing methods
1.	Blanching	1.	High-pressure processing (HPP)
2.	Pasteurization	2.	Pulsed electric fields (PEF)
3.	Sterilization	3.	UV light (UV)
4.	Drying	4.	Ultrasound (US)
5.	Chilling	5.	Ozonation
6.	Freezing	6.	Irradiation
7.	Refrigeration	7.	Pulse light
8.	Microwave		
9.	Ohmic heating		
10.	Infrared heating		

(A) Thermal Food Processing Mehods

1. Blanching

Blanching is a pre-heat treatment (unit operation) prior to freezing, canning, or drying in which fruits or vegetables are heated for the purpose of inactivating natural/endogenous enzymes; modifying texture; preserving color, flavor, and nutritional value; and removing trapped air.

Blanching is a process to scald foods in hot water.The process consists of three stages - preheating, blanching, and cooling.

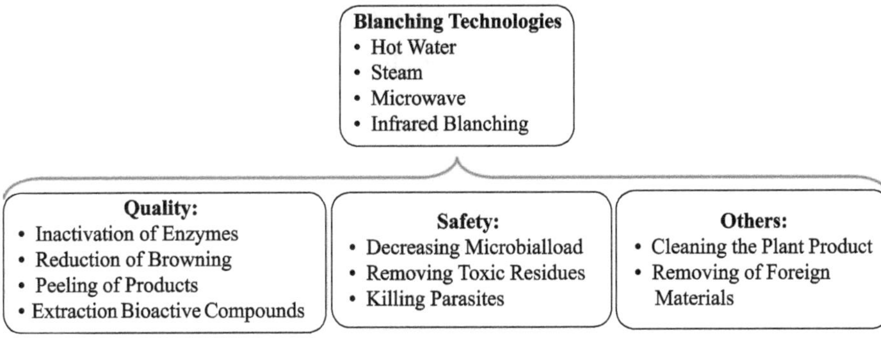

2. Pasteurization

- Pasteurization is a relatively mild heat treatment below 100°C (212°F) to inactivate pathogenic microorganism in packaged and non-packaged foods.
- In acidic foods (pH < 4.5, for example bottled fruit juice), it is used to extend the shelf life for several months by destruction of pathogenic and spoilage microorganism (yeasts and fungi) and enzyme inactivation (pectinesterase and polygalacturonase).
- Usually a liquid is heated to a specific temperature for a predefined time period followed by immediately cooling step (E.g. For fruit juices: The minimum processing condition has to be 63-66°C for 30 min; 77°C for 1 min; 88°C for 15s).

Fruit/vegetable source(s)	Product	Processing conditions	Key finding(s)
Amla, bael	Juice blend	80 to 90°C /25 s	Juice treated at 90°C showed best results for the nutritional quality of product
Apple	Juice	90°C/30 s	Complete inactivation of L. brevis and S. cerevisiae. 95.3% and 90.9% inactivation of PME and PPO

3. Sterilization

Sterilization is a controlled heating process used to completely eliminate all living micro-organisms, including thermoresistant spores in milk or other **food**. It can be achieved by moist heat, dry heat, filtration, irradiation, or by chemical methods.The food sterilization methods are divided into two categories: sterilization by heating (thermal processing) and sterilization without heating (non thermal processing).

Commercial Sterilization of canned foods

- Amongst the two groups of spore producing microorganisms *Clostridium* is more heat resistant than *Bacillus.* Temperatures of 110°C will kill most Bacillus spores within a short time. In the case of *Clostridium* temperatures of up to 121°C are needed to kill the spores within a relatively short time (HTST).

Sterilization process in the canned product can be subdivided into three phases.

- Phase 1 = heating phase: By means of a heating medium (water or steam) the product temperature is increased from ambient to the required sterilization temperature
- Phase 2 = holding phase: This temperature is maintained for a defined time
- Phase 3 = cooling phase: the temperature in the can is decreased by introduction of cold water into the autoclave.

Commercial sterility: fruit Juice or vegetable with a pH > 4.6 or above (low acid food) are given heat treatment in hermetically sealed canned to obtained commercial sterility (a 12-D process). Commercial sterility of thermally processed food" means the condition achieved by application of heat, sufficient, alone or in combination with other appropriate treatments, to render the food free from microorganisms capable of growing in the food at normal non-refrigerated conditions at which the food is likely to be held during distribution and storage.

4. Drying

- Drying refers to the removal of small amount of moisture from a solid or nearly solid material by evaporation to a predetermined level.
- Drying involves heat and mass transfer operations.
- Dehydration means complete removal of moisture content to bone dry condition. **Drying is based on heat input:**
- Direct (conduction)
- Indirect (convection)
- Radiant (radiation)

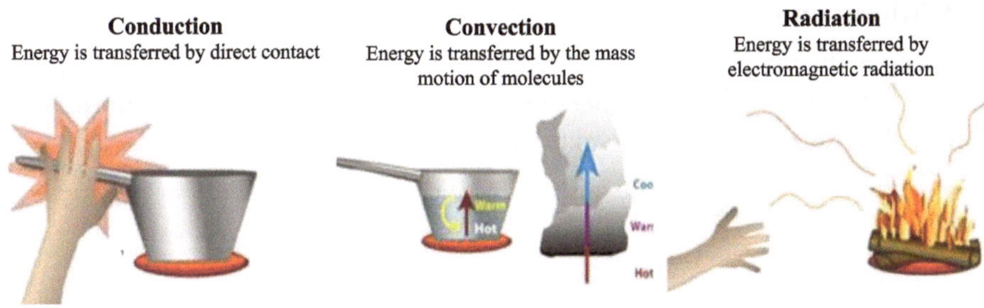

Objective of Drying

- To extend the shelf life of foods by reducing moisture content and water activity.
- To reduce the weight and volume thus ease of handling, reduce storage & transportation cost.
- To make food availability in lean season and improve palatability.
- Improve additional processing.
- Helps in minimizing post-harvest loses.
- Easy and economical way to save surplus food for a later time
- Provide an important modern marketing requirements

5. Preservation by Cold Treatment

- In contrast to heat treatment, low temperature practically does not destroy microorganisms or enzymes but merely depresses their activity. Therefore:
- Refrigeration retards spoilage but it cannot improve the initial quality of the product, hence the importance of assuring particularly high microbial quality in the starting material
- Unlike thermal sterilization, refrigeration is not a method of 'permanent preservation'
- Refrigerated and even frozen foods have a definite 'shelf life', the length of which depends on the storage temperature
- The preserving action of cold exists only as long as low temperature is maintained, hence the importance of maintaining a reliable 'cold chain' all along the commercial life of the product
- Refrigeration must often be combined with other preservation processes (the 'hurdle' principle).

6. Chilling and Freezing

Chilling is the application of low temperatures in the range of 0°C to 8°C, i.e. above the freezing point of the food, while freezing uses temperatures well below the freezing point, conventionally below −18°C.

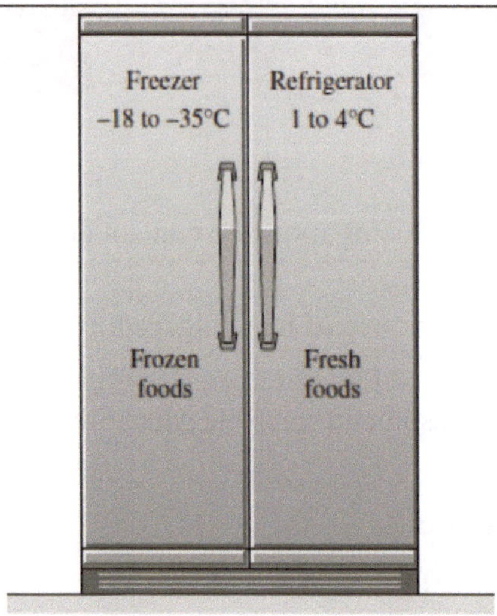

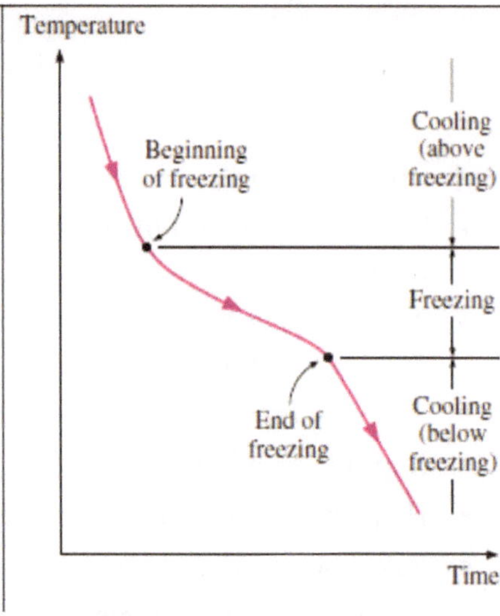

Slow freezing	Fast (Quick) freezing
• Rates of cooling of less than 1°C/min. Ice crystals form in extra-cellular locations • Large ice crystals formation • Maximum dislocation of water • Shrinkage (shrunk appearance of cells in frozen state) • Less than maximum attainable food quality	• Produces both extra-cellular and intracellular (mostly) locations of ice crystals • Small but numerous ice crystals • Minimum dislocation of ice crystals • Frozen appearance similar to the unfrozen state • Food quality usually superior to that attained by slow freezing

7. Refrigeration

The term refrigeration means cooling a space, substance or system to lower and/or maintain its temperature below the ambient one (while the removed heat is rejected at a higher temperature).

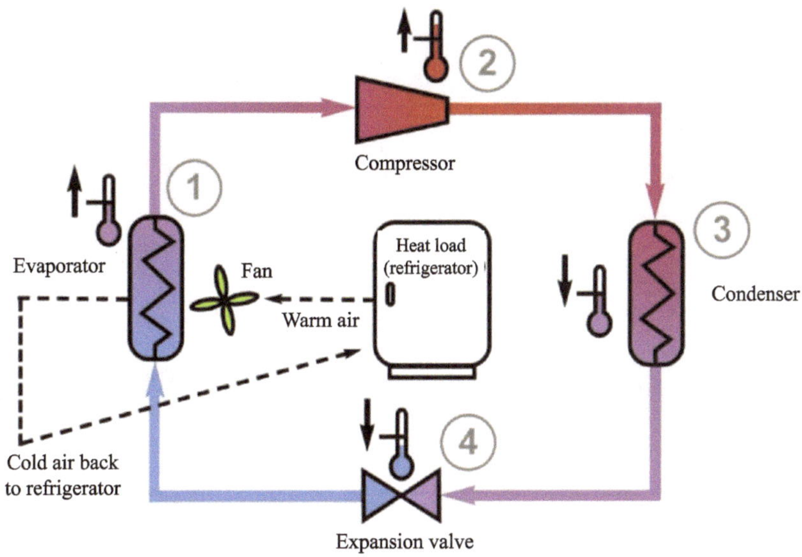

	Temperature	Storage time	Microbial activity
Refrigeration	16 to -2°C	Days or weeks	Below – 9.5°C no significant growth
Freezing	-18°C or below	Months or years	

Refrigeration installations may be divided into two groups

1. Devices the main purpose of which is to maintain the low temperature of chilled or frozen goods. Examples include cold storage facilities, refrigerated vehicles.
2. Devices the main purpose of which is to lower the temperature of food rapidly. Examples include chillers, freezers.

In contrast to heat treatment, low temperature does not destroy microorganisms or enzymes but merely depresses their activity. Refrigeration retards spoilage but it cannot improve the initial quality of the product, hence the importance of assuring particularly high microbial quality in the starting material.

8. Microwave heating

Microwave is an electromagnetic radiation of short wave length, reflected by metal, pass through the air, absorbed by several food constituents (dipolar H_2O).

Frequency approved for food application: 2450 MHz and 915 MHz

Microwave: Rapid heating rate (Volumetric heating) with short processing time

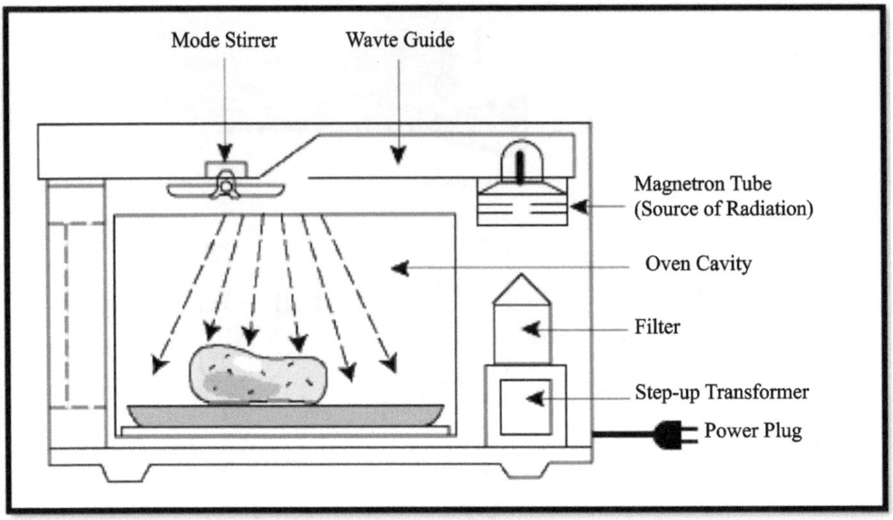

Schematic diagram of a microwave oven

Applications

- Microwave blanching
- Microwave pasteurization
- Microwave Sterilization
- Microwave Drying

Green bean blanched in microwave (Domestic 605W, 2450 MHz for 60 s) retained 88.5% Vitamin C compared to 84.7% in hot water blanching (100 °C for 2 min)

9. Ohmic heating

Ohmic heating is also called as electrical resistance heating or Joule heating in which an AC current is passed through a liquid particulate food which serve as an electrical resistance in which heat is generated.It provide rapid and uniform heating due to internal heat generation.

Rate of heating is directly proportional to the square of electric field strength (E) and electrical conductivity (σ). $P = \sigma V E^2$

Advantages:

- High energy efficiency (90% electrical energy is converted into heat)
- Rapid (1-10°C/s) and uniform heating
- Better nutrient and vitamins retention
- Reduced maintenance because of the lack of moving parts

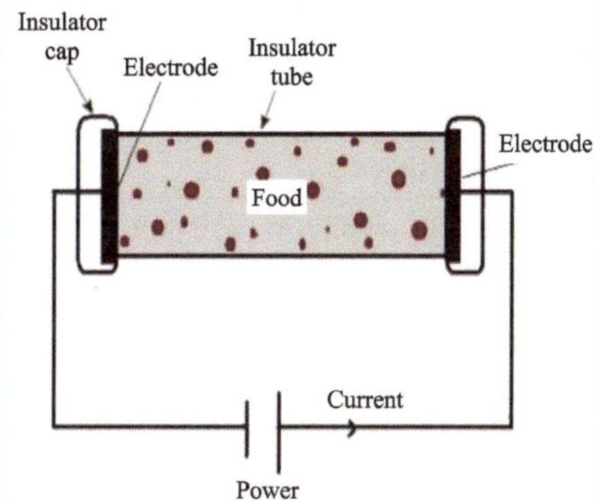

10. Infrared heating

IR radiation can be classified into 3 regions, namely, near-infrared (NIR), mid-infrared (MIR), and far-infrared (FIR), corresponding to the spectral ranges of 0.75 to 1.4, 1.4 to 3, and 3 to 1000 μm, respectively.By exposing an object to infrared (IR) **radiation(wavelength of 0.78 to 1000 μm- FIR region),** the heat energy generated can be absorbed by food materials.

- Infrared (IR) heating provides significant advantages over conventional heating, including reduced heating time, uniform heating, reduced quality losses
- Infrared heating can be applied to various food processing operations, namely, drying, baking, roasting, blanching, pasteurization, and sterilization.

(B) Non Thermal Processing of Food

Why non thermal processing?

- Thermal processing mainly aims at inactivation of pathogenic microorganisms and spores (depending on the treatment) to provide consumers with a safe product.
- However, despite the benefits of thermal treatment, the product quality may be altered due to undesirable changes that take place in the product with respect to flavour, colour, texture and general appearance.
- Consumers are becoming increasingly aware of the nutritional quality of food and the technologies used to process the food, showing a higher preference for fresh-like products free of chemicals and additives.

Non-thermal technologies were designed to eliminate the use of elevated temperatures during processing and to avoid the adverse effects of heat on the flavour, appearance and nutritive value of foods.

Novel non-thermal processes, such as high hydrostatic pressure (HHP), pulsed electric field (PEF), ionizing radiation (Irradiation) and ultrasonication, are able to inactivate microorganisms at ambient or sub-lethal temperatures.

Thermal methods	**Non thermal methods**
• Infrared radiation (0.76 - 350 μm) • Microwave radiation λ = 915 MHz or 2450 MHz; ν = 12.3 or 32.8 cm • Radio frequency (RF) radiation λ = 13.56, 27.12 MHz;ν = 7.4 - 22.1 m • Ohmic heating	• High pressure (100 - 800 MPa) • Pulsed electric field (0.5- 40 kV/cm) • Ultraviolet radiation (200 - 380 nm) • Ultrasonic waves (>20,000 Hz; 1.7 cm) • Ozonisation (gas or ozonated water) • Irradiation (Doses < 10 kGy) • Pulsed light (200 nm - 1 mm) • Membrane processing

1. High pressure processing (HPP)

- Typically, high pressure processing involves subjecting liquid or solid food to high hydrostatic pressures (100 to 800 MPa ; commercially, up to 600 MPa) for periods of a few minutes (1-20 min). The pressure is supplied in a uniform way and rapidly within the food material through a liquid phase, commonly water.
- The efficacy of high pressure processing is governed by **Le Chatelier's principle** stating that chemical reactions or physical processes associated with a decrease in molecular volume are favoured, whilst those accompanied with a volume increase are inhibited.

HPP treatments are independent of sample size and geometry, being mass/time independent, offering an interesting advantage compared to any heat treatment.

Advantages

Ensuring food safety by its action against pathogens

Limitations

Lower impact on low aw products

A HPP equipment consists of a vessel into the sample is subjected to pressure, generally transmitted by the utilization of a pressurization fluid (normally water), a pressure generation system and a parameters-handling device

HHP treatment cycle

1. Loading packaged products in the vessel
2. Vessel closure and filling with PTM
3. Pressurisation
4. Pressurisation release and vessel opening

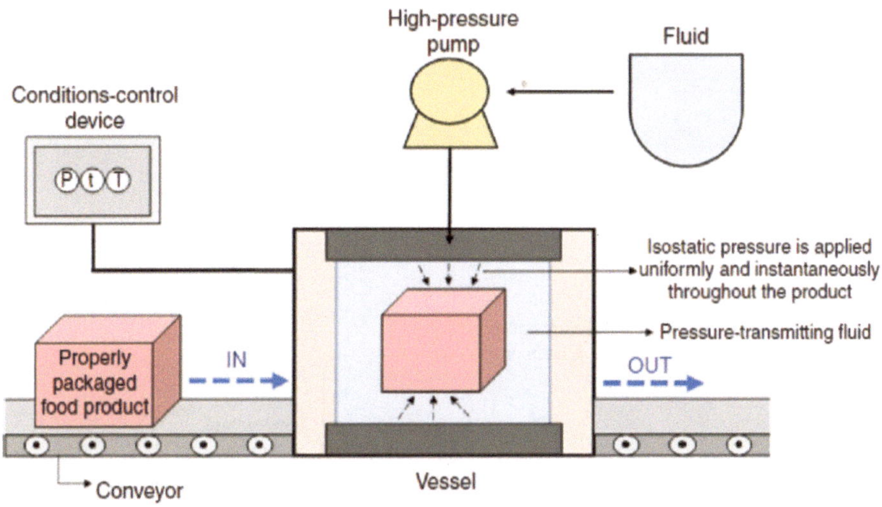

A schematic set up of High pressure processing (HPP)

2. Pulse Electric Field (PEF)

- High intensity pulse electric field involves the discharge of high voltage short electric pulses through the food product, resulting in mechanical effects on cell membrane without significant heating of the food .
- Electric fields in the range of 10–80 kV/cm generated by the application of short high voltage pulses (μs) between two electrodes achieve microbial inactivation at temperatures below those used in thermal processing (60–70°C).
- The mechanism by which microorganisms are inactivated by PEF is permeabilization of microbial membranes (cell electroporation).
- PEF technology enables inactivation of vegetative cells of bacteria and yeasts in various foods.

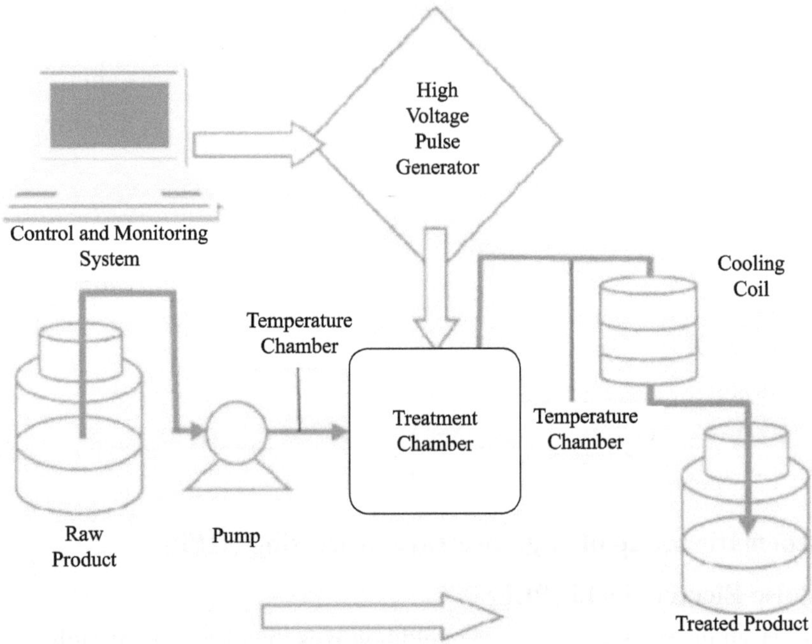

Flowchart of a PFE food processing system with basic component

As bacterial spores are resistant to PEF, applications of this technology mainly focus on food-borne pathogens and spoilage microorganisms, especially for acidic food products.

3. Ultraviolet (UV) Processing

Ultraviolet (UV) irradiation reduce the microbial load in liquid foods and surfaces, and to sterilize food packages and packaging materials, and environments involved in food processes.

UV light is subdivided into three regions as short-wave UV (UV-C, 200 and 280 nm), medium-wave UV (UV-B, 280 to 315 nm), and long wave UV (UV-A, 315 to 400 nm).

- The effectiveness of UV light on microorganisms results primarily from the fact that DNA molecules absorb UV-C photons between 200 and 300 nm, with peak absorption around 250-255 nm. This causes DNA damage by altering the nucleotide base pairing, thereby creating new linkages between adjacent nucleotides, particularly between pyrimidine bases, on the same DNA strand and ultimately results in cell death.

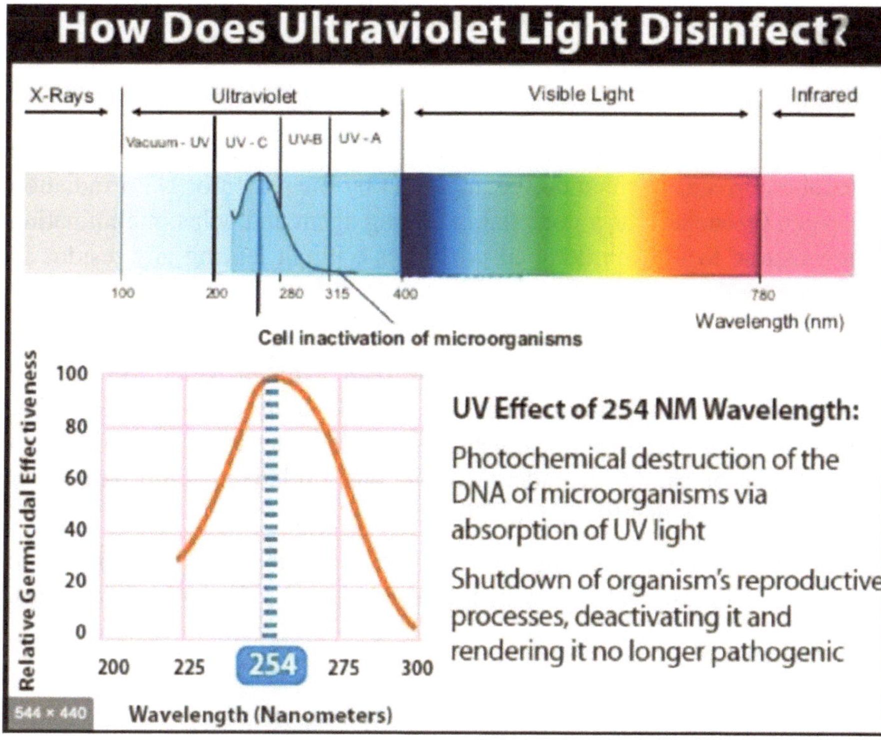

4. Ultrasound (US): Sonoprocessing

Ultrasound is a form of energy generated by sound waves of frequencies that are too high to be detected by human ear i.e., above 16 kHz.

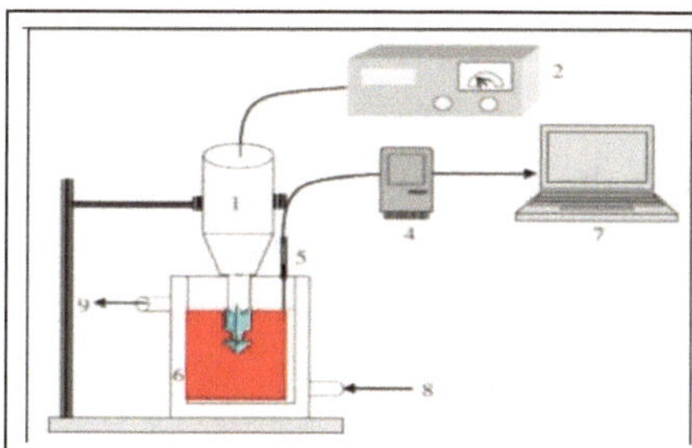

1. Ultrasound transducer
2. Ultrasonic generator
3. Ultrasound probe (19 mm)
4. Data logger
5. temperature probe
6. Jacketed beaker
7. Computer
8. Water inlet
9. Water outlet
10. Depth of probe in the sample (25 mm)

Low intensity ultrasound (at frequencies between 20 and 100 kHz and with intensities > 1 W/cm^2) is generally used in the food industry for analytical applications and to measure physicochemical properties of foods as they are non-destructive and do not cause changes in food properties. **Cavitation** is responsible for microbial inactivation. Best range of frequencies to achieve microbial inactivation is between 20 - 24 kHz. The inactivation of pathogenic microorganisms can be done using ultrasound alone or in combination with temperature (thermosonication)

5. Ozonation

Ozone (O_3) is an allotropic form of oxygen (O_2), formed naturally in the stratosphere in small amounts (0.05mg/litre) by the action of UV irradiation on oxygen.Ozone acts as a potential oxidizing agent and helps in eliminating pathogens due to its antimicrobial properties without leaving any residue in the treated food.

Potential utility of ozone to the food industry lies in the fact that ozone is 52% stronger than chlorine and effective over a much wider spectrum of microorganisms than chlorine and other disinfectants.ozone, unlike other disinfectants, leaves no chemical residues and degrades to molecular oxygen upon reaction or natural degradation.ozone has a relatively short half-life is both an asset and a liability to practitioners.

Fruits and vegetable washing :90% reduction of total bacterial counts upon treatment of Chinese cabbages with ozonated water (2.3 mg/L) for 60 minutes.

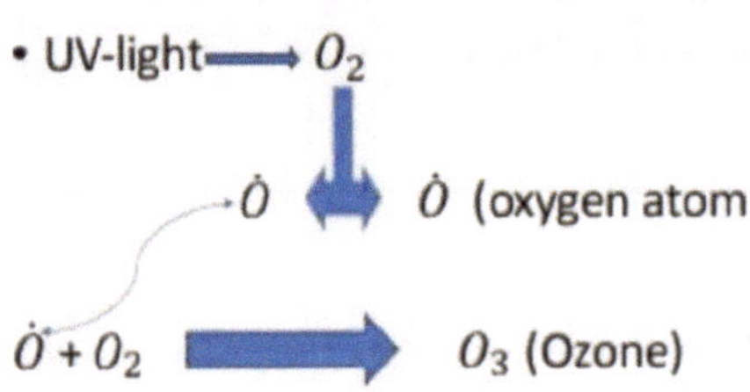

6. Irradiation of food

Food Irradiation: The process of exposing the food to controlled levels of ionizing radiation such as gamma rays to kill harmful microorganisms or to preserve its freshness.

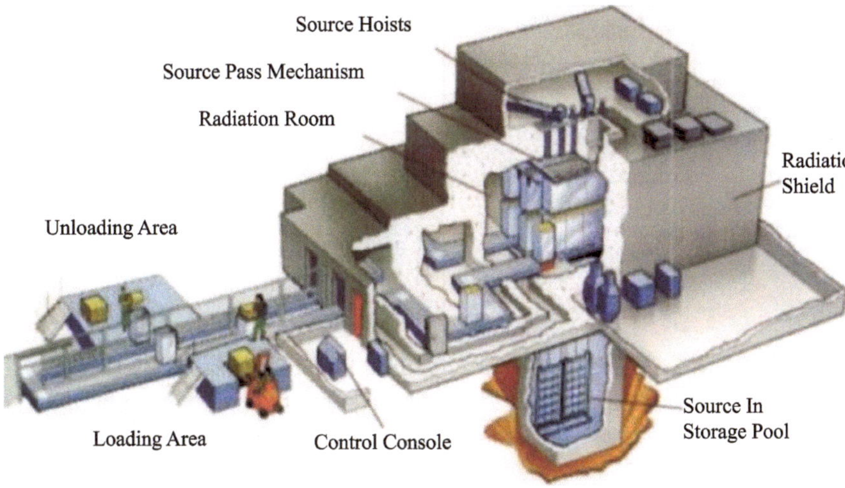

	The Radura is the International Symbol indicating the statement "treated with radiation" or "treated by radiation".

It is also called cold pasteurization. Gamma rays is produced by Cobalt-60, Cesium -137 (1- 2 Mev).

For fruits and vegetables: less than 1 kGy dose is used to inhibit sprouting (e.g onion, garlic and potato).

7. Pulse light processing

Pulse light: It involves the application of high intensity and short duration pulses (100-400 microseconds) of a continuous broad spectrum light in which around 40% of the light emitted corresponds to the UV region.

These pulses have high peak energy, producing a light intensity per time unit approximately 20,000 times greater than sunlight at sea level.

Components of a Pulse Light System	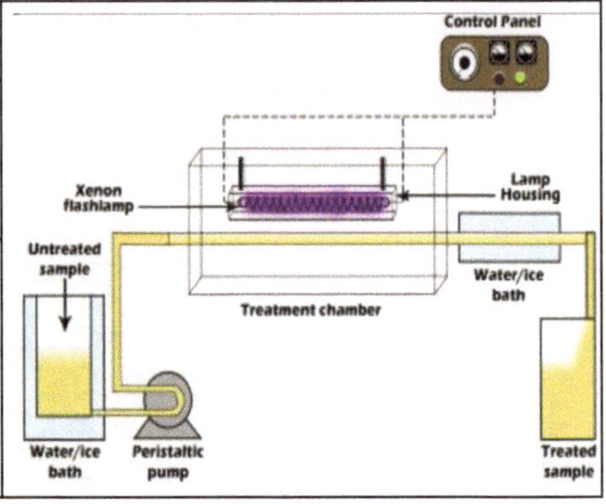
1. A pulse generator (comprising a high voltage DC power supply, capacitor and pulse controlling system)	
2. A treatment chamber (including flash lamp, lamp cooling system and reflectors).	

Fluences that range from 0.01-50 J/cm^2 are capable of inactivating pathogens and spoilage microorganisms, including viruses, molds and bacteria (vegetative and spore forms), using treatment times measured in seconds and without a detrimental increase in product temperature.

Conclusion

In terms of product quality and food safety biological and chemical activities must be brought under control and this is usually provided by traditional thermal processing.To provide thermal processes, enzymatic and microbiological inactivations in food are the most basic methods that are used. However, the height of the applied temperature has a negative impact on important parameters, which are the colour of food, nutritional value, sensory characteristics and flavour. Thermal treatment can kill microorganisms but also lead to physicochemical and sensory quality damage, especially to temperature-sensitive foods. Therefore, in recent years, producers and consumers are engaged in healthy food in the quest to obtain the quality parameters with less deviation. From this point of view non-thermal new techniques have emerged.The applications of novel non-thermal food processing technologies, including high pressure processing (HPP), ultrasound (US), pulsed electric fields (PEF), pulsed light (PL), and ozonation can extend the shelf-life by microbial inactivation and also keep good sensory quality attributes of different perishable food.

References

Barbosa-Cánovas, G. V., Pothakamury, U. R., Palou, E., Swanson, B. G., 1998. Food irradiation. Non thermal preservation of foods, pp.161-213. New York: Marcel Dekker.

Bintsis, T., Litopoulou-Tzanetaki, E. and Robinson, R.K. 2000. Existing and potential application of ultraviolet light in the food industry-a critical review. J. Sci. Food Agr. 90: 637-645

Bolton, J. R. and Linden, K. G., 2003. Standardization of methods for fluence (UV dose) determination in bench-scale UV experiments. J. Environ. Eng. 129: 209-215

Dunn, J., Ott, T. and Clark, W., 1995. Pulsed light treatment of food and packaging. Food Technologist. 49(9): 95–98

Fellows, P.J. 2017. "Food Processing Technology: Principles and Practice, 4th edition CRC Press, Boca Raton, FL

Holdsworth, D. and Simpson, R. 2008. Thermal Processing of Packaged Food, Springer, New York,

Singh, R. P. and Heldman, D.R., 2014. Introduction to Food Engineering, 5th Edition, Academic Press, NY,

Smelt, J.P.P.M., 1998. Recent advances in the microbiology of high pressure processing. Trends in Food Science & Technology, 9: 152-158

10

Fluidized Bed, Freeze and Spray Drying for Secondary Processing

A.K. Senapati and Dev Raj

Centre of Excellence on Post Harvest Technology, ASPEE College of Horticulture Navsari Agricultural University, Navsari, Gujarat-396 450, India

Introduction

Drying is one of the oldest methods of preserving food. Drying and dehydration both can be defined as the process of removing moisture from the foods. It is a simultaneous heat and mass transfer process, wherein heating medium or internal heat generation helps in evaporation of free water molecules from the product. Drying means removal of moisture of food upto a predetermined level where as in dehydration, complete removal of moisture upto bone dry weight of food material. Dehydrated foods have less than 2.5 % moisture content (dry basis) where as dried foods have dry basis moisture content more than 2.5 %. Drying operation consumes energy from 0.38 to 0.63 GJ/ton of food grain. Removal of water from foods using thermal energy thus makes it an energy intensive process. Modes of heat Transfer in various Driers are given in Table 1.Developments in dryer and methods, dryer selection based on form of raw material and typical residence time for various dryer are are given in Table 2,3 and 4, respectively.

Table 1: Mode of heat Transfer in various Driers

Heat Transfer Modes (Dryers)				
Convection	**Conduction**	**Radiation**	**Dielectric**	**Combined Modes**
Flash Dryer Spray Dryer Fluidized Bed Dryer Cabinet Dryer Tunnel Dryer Rotary Drier Spouted Bed Drier	Drum Drier Agitated pan Drier Rotary Drier Tray Drier	Infrared Shelf dryer Sun dryer	Microwave oven Microwave Tunnel Radio frequency Dryer	Microwave convective dryer Infrared convective dryer Microwave spouted dryer Radio frequency heat pump dryer

Table 2: Developments in Dryer and methods

Developments	Dryer / methods
1st generation dryers	Tunnel dryer, tray dryer, cabinet
2nd generation dryers	Spray dryer, drum dryer
3rd generation dryers	Freeze dryer, osmotic dehydration
4th generation dryers	Microwave, radio frequency, infra red, super heated steam, fluidized bed dryer, refractance window dryer

Table 3: Dryer Selection based on form of raw material

Nature of feed	Liquid			Cakes	Free flowing solids			
	Solution	**Slurry**	**Pastes**	**Centrifuge**	**Filter**	**Powder**	**Granule**	**Formed solids**
Convection dryer								
Belt conveyor dryer								
Fluidised bed dryer	√	√		√	√		√	√
Rotary dryer				√	√	√	√	
Spray dryer	√	√	√					
Tray dryer (batch)				√	√	√	√	√
Tray dryer (continuous)				√	√	√	√	
Conduction dryer								
Drum dryer								
Tray dryer (batch)				√	√	√	√	√

Tray dryer (continuous)				√	√	√	√	
Freeze dryer	√	√	√	√	√		√	√

Table 4: Typical Residence Time for Various Dryer

Dryers	**0-10 s**	**10-30 s**	**5-10 min**	**10-60 min**	**1-8 h**	**> 8 h**
Convective dryer						
Belt conveyor dryer				√		
Flash dryer	√					
Fluidised bed dryer				√		
Rotary dryer				√		
Spray dryer		√				
Tray dryer (batch)					√	
Tray dryer (continuous)				√		
Conduction dryer						
Drum dryer		√				
Steam jacked rotary dryer				√		
Freeze dryer						√
Radiation drying				√		
Microwave dyer				√		
Infra red dryer					√	

Fluidized bed dryers (FBD): Fluidized bed dryers are used extensively for the drying of wet particulate and granular materials that can be fluidized slurries, pastes and suspensions that can be fluidized in beds of inert solids. They are commonly used in processing many products such as chemicals, carbohydrates, foodstuff, biomaterials, beverage products, ceramics, pharmaceuticals in powder or agglomerated form, healthcare products, pesticides & agrochemicals, dyestuffs and pigments, detergents and surface-active agents, fertilizers, polymer and resins, tannins, combustion, incineration, waste management and environmental protection processes. Fluidized bed operation gives important advantages such as good solids mixing, high rates of heat and mass transfer, and easy material transport. For drying of powders in the particle size range of 50 to 2000 mm, fluidized beds compete successfully with other more traditional dryer types, e.g., rotary, tunnel, conveyor, continuous tray dryer, etc. Conventional fluidized bed is formed by passing a air stream from the bottom of a bed of particulate solids. At low air velocities, the bed is static (packed).The bed of particles rests on a air distributor plate. The fluidizing air passes through the distributor and it is uniformly distributed across the bed.

Pressure drop across the bed increases as the fluidizing velocity is increased. At a certain velocity, the bed is fluidized when the air stream totally supports the weight of the whole bed. This state is known as minimum fluidization and the corresponding air velocity is called minimum fluidization velocity, umf. Pressure drop across the bed remains nearly the same as pressure drop at minimum fluidization even if the gas velocity is increased further. Figure 1.1shows various regimes of the particulate bed from packed to bubbling bed when the gas velocity is increased. The graphs show the bed pressure drops and bed voidage under various regimes. A fluidized bed is operated at superficial gas velocities higher than the minimum fluidization velocity, umf, normally at 2-4 umf. The minimum fluidization velocity is typically obtained from experiments. There are several ways to determine the minimum fluidization velocity experimentally. It can also be estimated using various correlations.

Working Principle: Fluid bed dryers work on the principle of fluidization, where a material is converted from a static solid like state to a dynamic fluid like state. In this process, hot gas or air is introduced through a perforated distribution plate into the area holding the material. This hot gas pumps through the spaces between solid particles. As the velocity of the gas or air increases, the upward forces on the particles increase, causing them to equal the gravitational forces below (at terminal velocity). When the velocity of the air is greater than the settling velocity of granules, the granules remain partially suspended in the gas stream. After some time, a point of pressure is reached at which frictional drag in the particles is equal to the force of gravity. The granules rise in the container because of high velocity gas and fell back in a random motion. This condition is said to be fluidized state. In this process, heated air is blown up through the food particles to just suspend them in a gentle boiling motion. The moist air is then exhausted out from the top. It is used to dry grains, peas and other particulates.In a fluidized bed dryer, hot air is passed at high pressure through a perforated bottom of the container containing granules to be dried. In dryer, the food material is maintained suspended against gravity in an upward flowing air stream. The granules are lifted from the bottom and suspended in the stream of air. Hot gas surrounding every granules completely dries them in an uniform manner. Uniform conditions of temperature, composition and particle size distribution is maintained. The direct contact between particles and air/gas is possible. Various regimes of a bed of particles at different air velocities are illustrated in Fig.1.

Construction of a vertical fluidized bed dryer: The fluidized bed dryer is made of stainless steel. It can be designed in either batch or continuous type fluid bed dryer. A detachable bowl is placed at the bottom of the dryer, which is used for charging and discharging (product container). The bowl has a perforated bottom with a wire mesh support for placing materials to be dried. A fan is mounted in the upper part for circulating hot air. Fresh air inlet, pre-filter and heat exchanger are connected serially to heat the air to the required temperatures. The temperature of hot air and exit are monitored. Bag filters are placed above the drying bowl for recovery of fines.

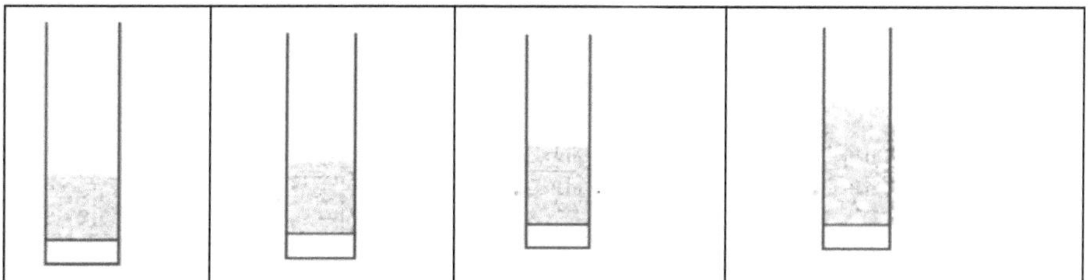

A= Fixed Bed	B=Expanded bed	C=Minimum Fluidization	D=Bubbling Fluidization

Fig. 1. Various regimes of a bed of particles at different air velocities.

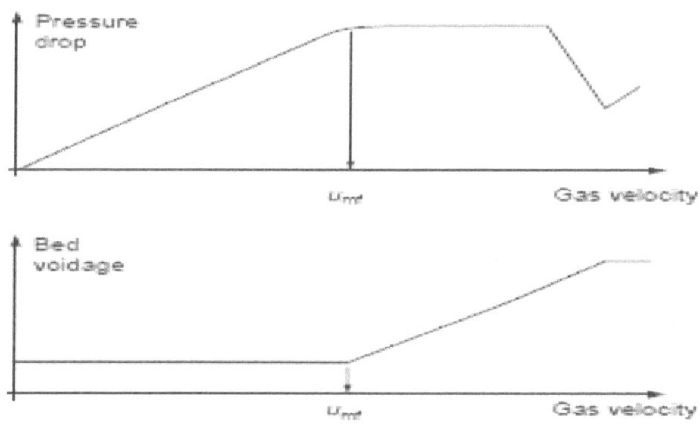

Fig. 2. Relation of Pressure during FBD

Advantages: The major advantages of fluidized bed drying include: Fluidized bed dryer requires less time to complete drying. i.e., 20 to 40 min. Hot spots are not observed in the dryer, because of its excellent mixing and drying capacities. The thermal efficiency is 2 to 6 times greater than tray dryer. It facilitates the drying of thermolabile substances, since the contact time for drying is short. It can be used either as batch type or continuous type. The free movement of individual particles eliminates the risk of soluble material migrating as may occur in static bed high rate of moisture removal, high thermal efficiency, easy material transport inside dryer, ease of control, and low maintenance cost.

Limitations: Limitations of fluidized bed dryer include high pressure drop, high electrical power consumption, poor fluidization quality of some particulate products, non uniform product quality for certain types of fluidized bed dryers ,erosion of pipes an d vessels, entrainment of fine particles, attrition or pulverization of particles , agglomeration of fine particles , etc. The general process flow of a fluidized bed drying is illustrated herewith Fig. 3.

Raw material collection

↓

Sample preparation includes Sorting and Grading

↓

Washing

↓

Pre treatment if required

↓

Final samples preparation prior to drying

↓

Drying of products under Fluidized bed drier

↓

Packaging and Labeling

↓

Storage

Fig. 3: The General Process Flow of a Fluidized Bed Drying

Application: Besides drying, fluidized bed has found wide ranges of industrial applications in various industries for mixing, de dusting, granulation, coating, agglomeration, cooling, and chemical reactions, incineration, combustion, gasification, etc. Many of these processes can be incorporated with fluidized bed drying in one unit process or to accomplish two or more processes in the same unit. Processes that can be advantageously incorporated with fluidized bed drying. The mixing effect in a fluidized be d is generally good for particle sizes between 50 and 2000 mm. For fine particles (particle size less than 50 mm), or for particles that are difficult to fluidize when wet, vibration is normally app lied to improve the fluidization quality and the mixing effect. For large particles, insertion of internals or use of the spouting mode can help to improve the operation. For fluidized bed drying, good particle mixing is essential. Thus, knowledge on particle fluidization characteristics and their properties is required to ensure good performance of a fluidized bed dryer. In addition, the bed of particles can be fluidized by a pulsating flow or by fluidizing sections of the bed periodically such that the entire bed is fluidized in sequence once over a cycle. Clearly, this operation results in saving of drying air and hence electrical power but it also leads to a longer operating time due to the intermittent mode of heat input. Besides, intermittent fluidization can reduce problem of mechanical damage to the particles due to continuous vigorous particle–particle collision as well as attrition-induced dusting.

Spray drying: This technique is used for drying of liquid foods to convert them to powder. The liquid food is passed through an atomizer (at a temperature of about 200°C), which breaks the food into minute droplets, resulting in drying of material in seconds. This method of dehydration can produce exceptionally high quality dried products. Spray drying is a suspended particle processing (SPP) technique that utilizes liquid atomization to create droplets that are dried to individual particles when moved in a hot gaseous drying medium, usually air. It is a one step continuous unit processing operation. It has become one of the most important methods for drying the fluid foods. The development of the process has been intimately associated with the dairy industry and the demand for drying of milk powders. Spray drying used in dairy industry dates back to around 1800, but it was not until 1850 that it became possible in industrial scale to dry the milk. However, this technology has been developed and expanded to cover a large food group that is now successfully spray dried. Spray dryers are now estimated to be commercially in use to dry products from agrochemical, biotechnology products, fine and heavy chemicals, dairy products, dyestuffs, mineral concentrates to pharmaceuticals in capacities ranging from a few kg/h to over 50 tons/h evaporation capacity. The general process flow of a spray drying is illustrated herewith Fig. 4.

Raw material collection

↓

Sample preparation includes Sorting and Grading

↓

Washing

↓

Pre treatment if required

↓

Final samples preparation(Juice extraction) prior to drying

↓

Drying of products under spray drier

↓

Dried Powder

↓

Packaging and Storage

Fig. 4: The General Process Flow of a Spray Drying.

Working Principle: The spray drying process operates in the following way, i.e., the liquid is pumped from the product feed tank to the atomization device, such as the rotary disc atomizer, pressure nozzle, pneumatic nozzle, and ultrasonic nozzle, which is usually located in the air distributor at the top of the drying chamber. The drying air is drawn from the atmosphere through a filter by a supply fan and is passed through the air heater, e.g., oil furnace, electrical heater, steam heater, etc., to the air distributor. The droplets produced by the atomizer meet the hot air and the evaporation takes place cooling the air in the meantime. After the drying of the droplets in the chamber, the majority of the dried product falls to the bottom of the chamber and entrains in the air. Then they pass through the cyclone for separation of the dried particles from the air. The particles leave the cyclone at the bottom via a rotary valve and are collected or packed later. The fine particles will remain entrained in the air and then pass the air through scrubbers for washing out the fine particles from the air. The cycled washing liquid can be transported to the pretreated process of the feed. The air passes from the scrubber to the atmosphere via the exhaust fan. The process control system may comprise the necessary indication of the temperatures at the air inlet and outlet, the voltages and currents of the motors, etc. The automatic controls often are to maintain the inlet temperature by altering the heater operating parameters, such as the steam pressure, the amount of oil to the heater, etc., and the outlet temperature by adjusting the amount of feed pumped to the atomizer. Typical layouts of spray dryer unit with various parts are illustrated in Fig. 5.

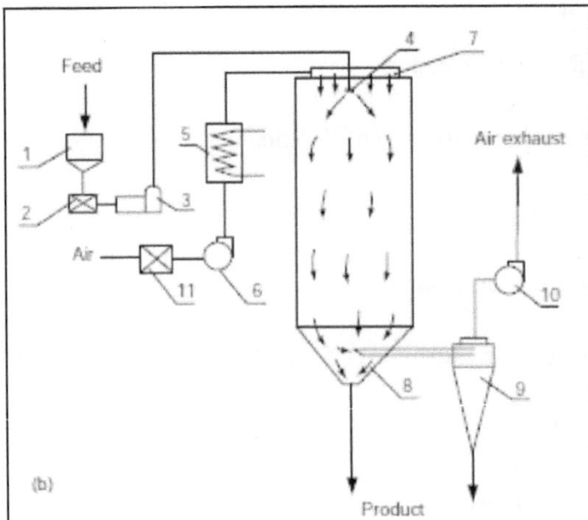

b. With nozzle atomizer

1. Feed tank
2. Filter
3. Pump
4. Atomizer
5. Air heater
6. Fan
7. Air disperser
8. Drying chamber
9. Cyclone
10. Exhaust fan

Fig. 5: Typical layout of Spray dryer Unit

Different parts of Spray Dryer

Advantages: The major advantages are that this technique can handle heat sensitive, non heat sensitive and heat resistant pumpable fluids as feed stocks from which a powder is produced, produce dry material of controllable particle size, shape, form, moisture content, and other specific properties irrespective of dryer capacity and heat sensitivity. It provides continuous operation adaptable to both conventional and PLC control. It handles wide range of production rates, i.e., any individual capacity requirement can be designed by spray dryers. Provide extensive flexibility in spray dryer design, such as drying of organic solvent-based feed stocks without explosion and fire risk; drying of aqueous feed stocks (where the resulting powders exhibit potentially explosive properties as a powder cloud in air); drying of toxic materials; drying of feed stocks that require handling in aseptic and hygienic drying conditions; drying of liquid feed stocks to granular, agglomerated and non agglomerated products.

Drawback: There are some limitations, such as high installation costs, lower thermal efficiency and product deposit on the drying chamber may lead to degraded product or even fire hazard.

Application: Spray dried products on industrial scale in food industry such as e.g., milk, whey, egg, soya protein, etc. Fruits and vegetables products like banana, tomato, coconut milk, etc. Carbohydrates, e.g., glucose, total sugar; Chemical industry like phenol formaldehyde resin, catalysts, PVC emulsion type, amino acids, etc. Ceramic industry like e.g., aluminum oxide, carbides, iron oxide, kaolin, etc. Dyestuffs and pigments like chrome yellow, food color, titanium dioxide, paint pigments, etc. Fertilizers like nitrates, ammonium salts, phosphates, etc. Detergent and surface active agents like detergent enzymes, bleach powder, emulsifying agents, etc.

Freeze drying: Freeze drying utilizes the principle that under high vacuum (27-133 Pa pressure), frozen water can be removed from a food and collected without going through a liquid phase. Because the material remains frozen, no damage because of heat occurs. In addition, there is little or no loss in sensory qualities of the product. Further, because the removal of ice crystals leaves a porous honeycomb structure, the products tend to rehydrate rapidly. However, freeze drying is slow and expensive. The long processing time requires additional energy to run the compressor and refrigeration unit, which makes the process very expensive for commercial use. Therefore, freeze drying is most often used for products that can either be sold at a premium prize or which can withstand only a small amount of sensory deterioration. In this method, the material such as fruit juice concentrates is first poured on trays in the lower chamber of a freeze drier, frozen and then dried in the upper chamber under high vacuum. The material is directly dried by sublimation of ice without passing through intermediate liquid stage. The dried product is highly hygroscopic. It reconstitutes easily. Mango pulp, orange juice concentrate, passion fruit juice and guava pulp have been prepared to give freeze-dried powders of excellent quality for taste, flavour and reconstitution property. Certain biological materials, pharmaceuticals, and foodstuffs, which may not be heated even to moderate temperatures in ordinary drying, may be freeze dried.

Working Principles: The substance to be dried is usually frozen conditions. In freeze drying, the water or another solvent is removed as a vapor by sublimation from the frozen material in a vacuum chamber. After the solvent sublimes to a vapor, it is removed from the drying chamber where the drying process occurs. As a rule, freeze drying produces the highest quality food product obtainable by any drying method. A prominent factor is the structural rigidity afforded by the frozen substance at the surface where sublimation occurs. This rigidity to a large extent prevents collapse of the solid matrix remaining after drying. The result is a porous, non shrunken structure in the dried product that facilitates rapid and almost complete rehydration when water is added to the substance at a later time. Freeze drying of food and biological materials also has the advantage of little loss of flavor and aroma. The low processing temperatures, the relative absence of liquid water, and the rapid transition of any local region of the material dried from a fully hydrated to a nearly completely dehydrated state minimize the degradative reactions that normally occur in ordinary drying processes, such as non enzymatic browning, protein denaturation, and enzymatic reactions. In any food material, some non frozen water, which is called bound or sorbed water, will almost unavoidably be present during freeze drying, but there is very often a rather sharp transition temperature for the still wet region during drying below which the product quality improves markedly. This improvement shows that sufficient water is frozen to give the beneficial product characteristics of freeze drying. Freeze drying requires very low pressures or high vacuum to produce a satisfactory drying rate. If the water was in a pure state, freeze drying at or near -8°C at an absolute pressure of 4.58 mm Hg could be performed. But, since the water usually exists in a combined state or a solution, the material must be cooled below -8°C to keep the water in the solid phase. Most freeze drying is done at -10°C or lower at absolute pressures of about 2 mmHg or less. In short, freeze drying is a multiple operation in which the material to be stabilized is Frozen hard by low temperature cooling. Dried by direct sublimation of the frozen solvent and by desorption of the sorbed or bound solvent (nonfrozen solvent), generally under reduced pressure. It is Stored in the dry state under controlled conditions (free of oxygen and water vapor and usually in airtight, opaque containers filled with inert dry gas). Typical layout of pilot scale freeze dryer with Various parts are illustrated in Fig. 6. General process flow chart for preparation of freeze dried products is given in Fig. 7.

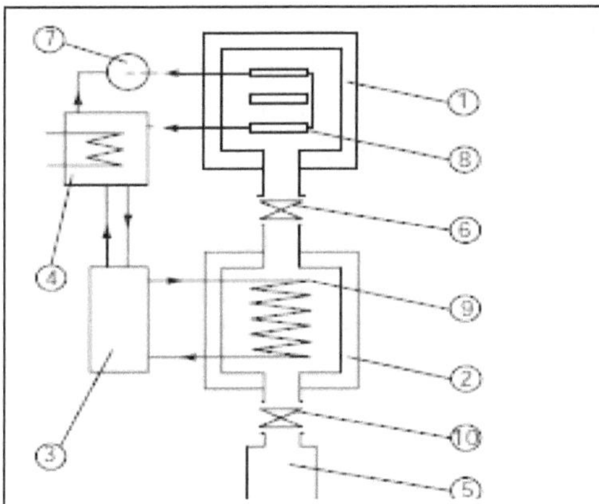 	1. Drying chamber 2. Ice condenser chamber 3. Refrigeration unit 4. Cooling and heating system for the plates 5. Vacuum unit 6. Isolation butterfly valve 7. Silicon oil pump 8. Cooling and heating plate 9. Refrigerated coil 10. Condenser vacuum valve
Fig. 6: Typical layout of Pilot scale freeze dryer	Various part of Freeze dryer

Collection of Raw material

↓

Washing and Sorting

↓

Slicing into various thickness as per standard dimension

↓

Blanching if needed

↓

Freezing -20°C

↓

Dried slices under freeze drier of products under spray drier

↓

Dried Powder

↓

Packaging, labeling and Storage

Fig. 7. General Process Flow Chart for Preparation of freeze dried products

Application: Freeze drying is mainly applied to different categories of materials such as blood plasma, serum, hormone solutions, foodstuffs, pharmaceuticals (e.g., antibiotics), ceramics, superconducting materials, etc.

Advantages: most of the processed products can be kept in retaining all their initial physical, chemical, biological, and organoleptic properties and products remaining available at any time for immediate better reconstitution.

Conclusion

Drying technology can be adopted as per the requirement of product quality, quantity, investment capacity and brand Image. The application of fluidized bed dryer in industries possesses much importance because of increase energy efficiency and operational efficiency. Fluidization gives higher benefit as compare to traditional method as it provides maximum surface area for the higher heat and mass transfer during drying. The economics of the process indicate that freeze drying can be suitable for high value products with specific biological or physico chemical properties. Freeze dried products can be prepared by suitable sizes of the final products , freezing at (-20°C) followed by freeze drying under -760 mm of Hg vacuum at 50°C drying temperature gives better quality of the finished products without altering quality. Spray drying process is one of the most common methods increasing applications in various industries but almost all industries need or use or produce powders starting from liquid products.

References

Abuaf, N. and Staub, F.W. 1986. Drying of liquid solid slurry droplets, Drying 86, Vol. 1, Ed. A.S. Mujumdar, Hemisphere/McGraw-Hill, New York, pp. 277-284.

Amelot, M.P. and Gauvin, W.H. 1986. Spray drying with plasma-heated water vapor, Drying '86, Vol. 1, Ed. A.S.Mujumdar, Hemisphere/McGraw-Hill, New York, pp. 285-290.

Gupta, C.K. and Sathiyamoorthy, D. 1999. Fluid Bed Technology in Material Processing, CRC Press, New York, chap.1.

Geldart, D. 1973. Types of gas fluidization, Powder Technol., 7: 285-292.

Liapis, A.I. 1987. Freeze drying. In: Handbook of Industrial Drying, 1st ed. (AS Mujumdar, Ed.), New York: Marcel Dekker, pp. 295-326.

Mellor, J.D. 1978. Fundamentals of Freeze Drying. London: Academic Press.

Mujumdar, A.S. and Devahastin, S. 2003. Applications for fluidized bed drying, in Handbook of Fluidization and Fluid Systems, Yang, W.C., Ed., Marcel Dekker, New York, chap. 18.

Mujumdar, A. S. 2006. Handbook of Industrial Drying. CRC, Lodz, Poland, chap. 8 and 10.

Vreedenberg, H.A. 1958. Heat transfer between a fluidized bed and a horizontal tube, Chem. Eng. Sci., 9(1): 52-60.

11

Osmotic Drying and Vacuum Frying

Nilam V. Patel and Dev Raj

Department of Post Harvest Technology, ASPEE College of Horticulture Navsari Agricultural University, Navsari, Gujarat-396 450, India

Introduction

The drying of fruits and vegetables during the season of abundance in order to preserve them for consumption during season of shortage is an ancient practice. Thus, drying is one of the oldest methods of preservation of the food commodities. It is practiced from ancient times and is a very common method of preservation in the tropical countries like India where conditions are much more favorable for sun drying. All micro-organisms need moisture to grow. So the removal of water in the food is brought down below a certain level at which they are unable to grow. Moisture content from food can be removed by different methods like sun drying (very oldest method and most popular), solar drying, cabinet drying, mechanical drying, tunnel drying, poly tunnel drying, adiabatic drying, spray drying, osmotic drying, fluidized bed drying etc. Grains are mostly dried in the sun. The moisture content of the grains can be reduced to a limit that is safer for their storage by sun drying. Many fruits and vegetables can also be dried by using sun energy. Some coastal areas of the countries fishes are also dried by exposing to the sun. After drying, sometimes the dried products are converted into powder. However, these powders are very hygroscopic and should be packed in airtight. Mango powder is prepared by this technique, but expensive and not popular in India. Nowadays many artificial methods have been developed for removal of the moisture. The relationship between the moisture content and temperature of air during process is referred as Psychrometric relation. An instrument used to measure the humidity (Relative humidity) of the atmosphere is known as Psychrometer (Hygrometer). The residual moisture in vegetable should not be more than 6-8% and in fruits it should be in the range of 10-20%. The dried fruit can be used as such or after soaking while dried vegetable are usually soaked in water overnight and then cooked. Vegetable are considered to be dry when they become brittle. There are generally two drying methods for foods like

1) Natural drying and 2) Artificial drying. The preservation / stability of the dehydrated products is conferred by the abstraction of water, so that the product is unable to support the microbial growth, its chemical forms of deterioration is inhibited and the drastic reduction in the weight / volume facilitate easier handling and marketing. Among dehydrated products, dehydrated onions have largest share of 45%. India is the third largest exporter of the onions in the world next to the Netherland and Spain. Besides this, India exports dehydrated bell pepper, garlic and mushrooms.

Drying is the removal of water under the influence non-conventional energy sources like sun and wind. A synonym of the drying is desiccation. Drying operation for fruit and vegetables is a complex one since this involves simultaneous exchange of moisture and heat. It is natural drying method. Dehydration is the removal of water / moisture by the application of artificial heat under controlled condition of temperature, humidity and air flow. It is artificial drying method (Fig 1).

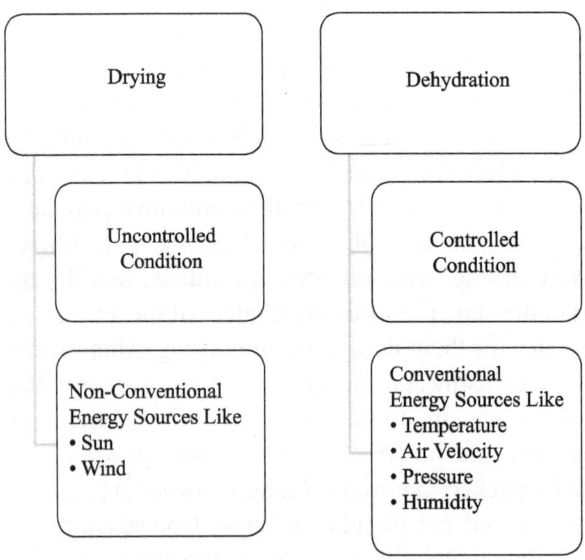

Fig. 1. Drying verses dehydration

Principle of Drying / Dehydration

The main principle of the dehydration is preservation of food by removal of the moisture from the food product with the application of heat or change of pressure. The phenomenon of heat and mass transfer is illustrated through Fig 2. So during drying or dehydration there is heat transfer in and mass transfer out.

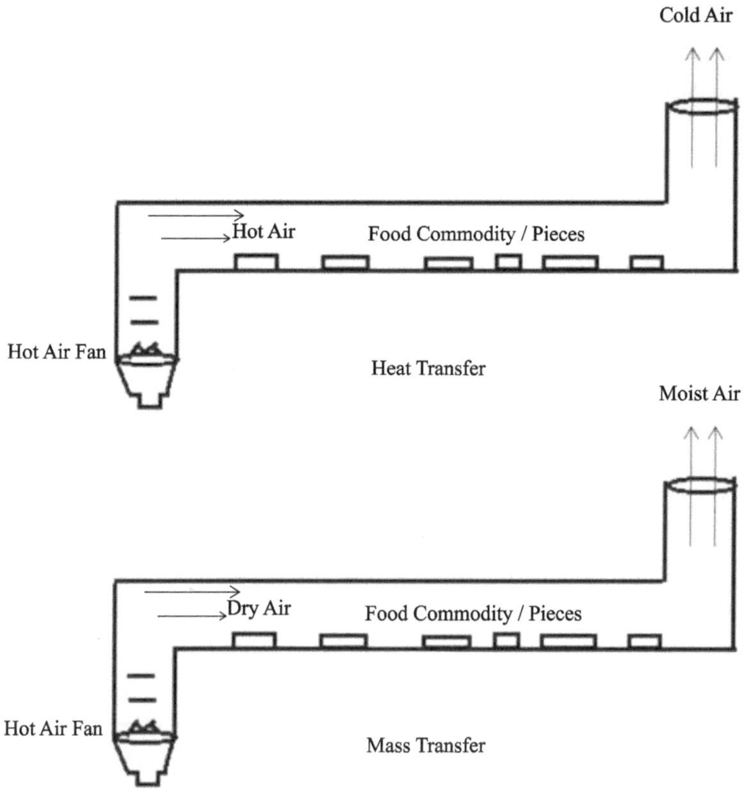

Fig. 2. Phenomenon of heat and mass transfer during drying of food

Mode of Preservation

All microbe needs moisture to grow in the food in which they harbor. The growth depends of all microbes depends on water activity present in the foods. The dehydration reduces water activity of foods to alarm water level and the reduced water activity results decreased spoilage of food via microbial deterioration and chemical changes leading to preservation of the food.

Objectives of Drying

- To increase the shelf life of the food products
- To maintain the quality characteristics of the dehydrated products
- To help in easy handling, storage and transportation.
- To reduce the bulk of the dried products.
- To prepare instant value added products.
- To find out suitable varieties for dehydration purpose to get higher yield.

Drying Methods

There are different drying methods like osmotic drying, Drum drying, spray drying, vaccum shelf drying, vacuum belt drying, atmosphere drying, kiln drying, tunnel drying, sonic drying, ball drying, electromagnetic drying, foam mat drying, puff drying, slush drying, pneumatic drying, azeotropic drying, microwave drying, hot oil immersion, thin process drying etc. For different drying methods, different mechanical dehydrators as discussed earlier are being used depending upon the commodity and drying techniques.

Osmotic Drying

It is removal of the moisture from the food by the addition of humectants (glycerol, sugar etc) which exert osmotic action on the microorganisms by binding free water and make it unavailable to microbes and resulting in increase in non-refrigerated shelf life. In this dehydration, the water diffuses from dilute solution to concentrated solution (hypertonic solution) through a semi-permeable membrane until concentration equilibrium is reach. In this technique there is simultaneous counter current mass transfer of water from the solution to hypertonic solution.

Thus, it is a process that involves the partial removal of water from food items such as fruits and vegetables. The process is based on a tendency to reach equilibrium between osmotic pressure inside the biological cells and the surrounding osmotic solution, which has an increased osmotic pressure caused by high concentration of soluble osmotic agent. Molecular diffusion of water through semi-permeable membranes takes place and product loses its water and increase soluble solids content. Osmosis is spontaneous movement of a solvent (such as water) from a lower-concentration to a higher-concentration solution, through a semi-permeable membrane. The flow of solvent stops when both solutions become equal in concentrations (Fig 3). The mechanisms of osmotic dehydration is illustrated through Fig 4.

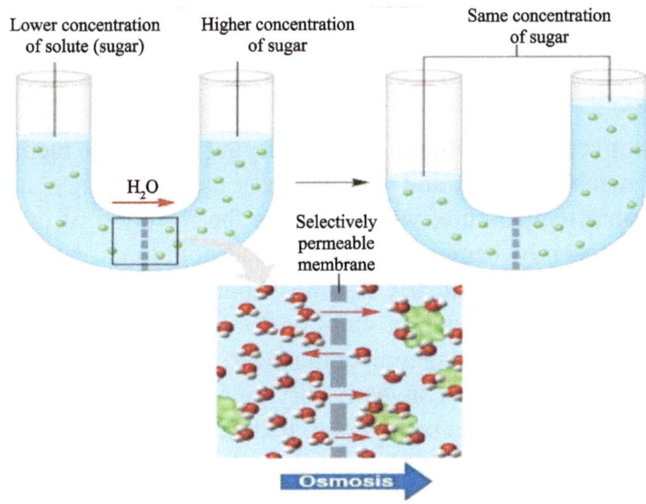

Fig. 3. Principle of Osmosis

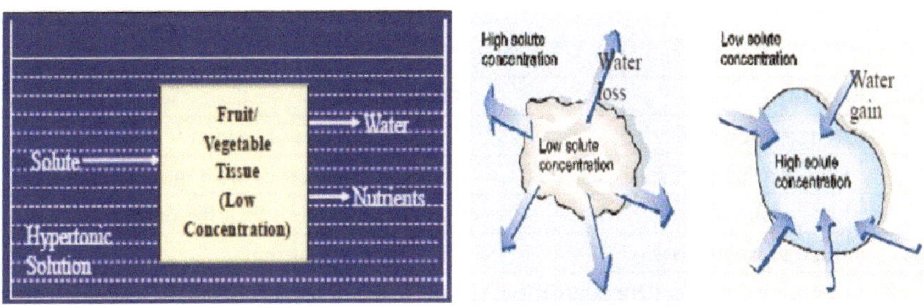

(a) Mass transfer across a fruit/vegetable tissue during osmotic dehydration

(b) Effect of high and low solute concentration inside and outside the cell on mass transfer

Fig. 4. Mechanisms of osmotic dehydration and effect of concentration

Different osmotic agents

1. Calcium Chloride
2. Ethanol
3. Fructose
4. Invert Sugar
5. Lactose
6. Malto Dextrine
7. Sodium Chloride/salt
8. Sucrose/Sugar
9. Starch/Corn syrup

Objectives of osmotic drying

1. Quality Improvement and
2. Energy Saving

Comparing to other conventional methods, osmotic drying is a simple method which requires less mechanical aid and involves less electric energy. It is easy to perform at room temperature, which ensures the retention of color, texture and nutrients. Recently, osmotic drying of food got attention due to its importance in food processing industry. The most commonly used osmotic agents are sucrose for fruits and sodium chloride for vegetables. Osmotic drying can be conducted without raising temperatures *i.e.* ambient temperature hence, it is less energy intensive than convective drying or freezing. The main advantages of osmotic drying include better color, texture and flavor retention due to minimum heat damage. Osmotic pretreatment *viz.,* blanching, to solid foods is used to improve the nutritional, sensorial and functional properties of food without changing its integrity.

Difference between Drying v/s Osmo-drying

S.No.	Drying	Osmo-drying
1	Rehydration is needed.	Rehydration is not needed.
2	Sensorial properties are affected.	Sensorial properties are v. good.
3	Loss of flavor.	Retention of fresh fruit flavors.
4	Huge nutrients loss.	Retention of nutrients.
5	Chance of physical contamination.	No chance of contamination.

*Hence, for improved quality & Ready to Eat products osmo-drying is the best choice!!!

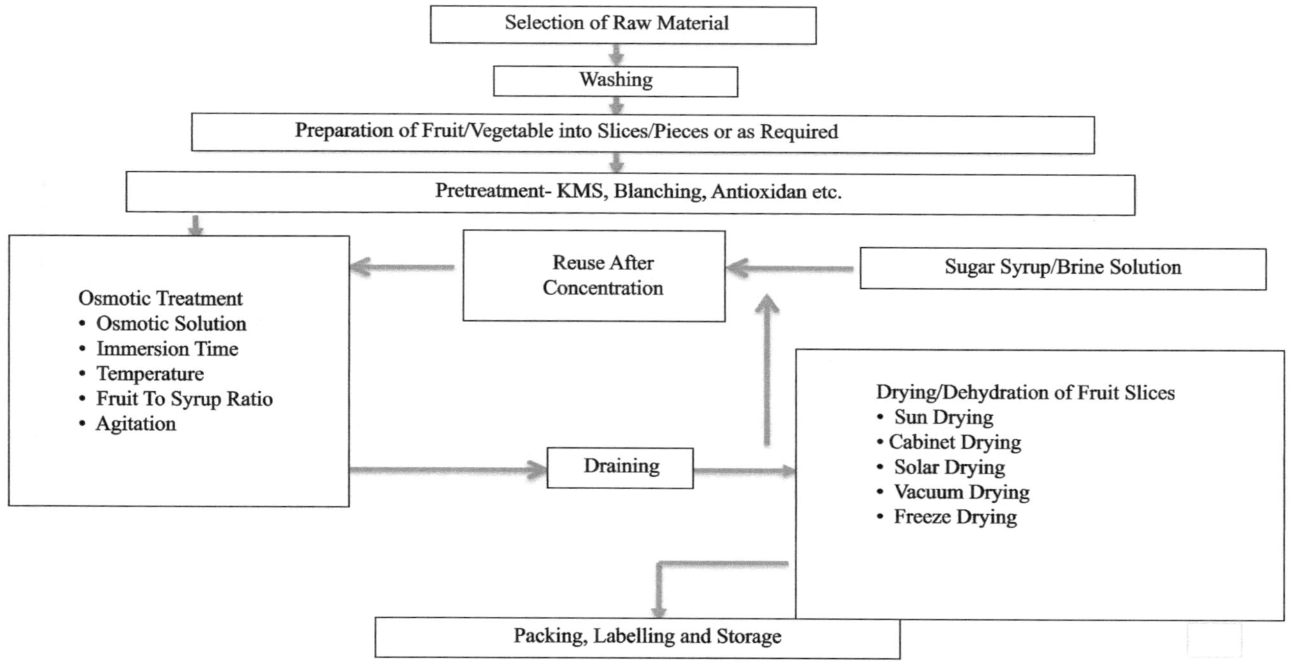

Fig. 4. Flow chart of general method of osmotic drying / Dehydration

Osmotic drying of fruits and vegetables

- Syruping is a preservation method in which sugar is used as a preservative that reduces browning by preventing the entry of oxygen.
- Brining is one of the oldest food preservation method. It is an inexpensive and easy way to preserve vegetables.
- It is a process where the common salt (NaCl) is used as a preservative that penetrates into the tissue; hence slows the bacterial growth and deactivates the enzymes. Salted or brined vegetables retain a fair amount of their nutritive value.
- This is a two steps of preserving fruits and vegetables:-
    - Dipping of the fruits in concentrated sugar syrup/ brine solution (osmosis).
    - Final drying under the sun or use of dryers (Dehydration).

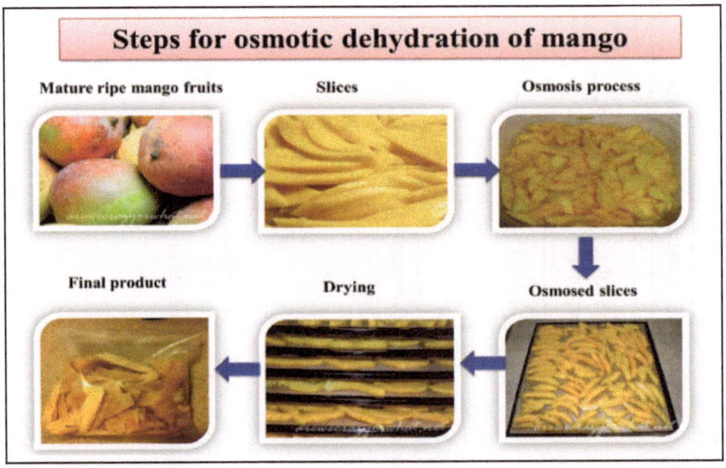

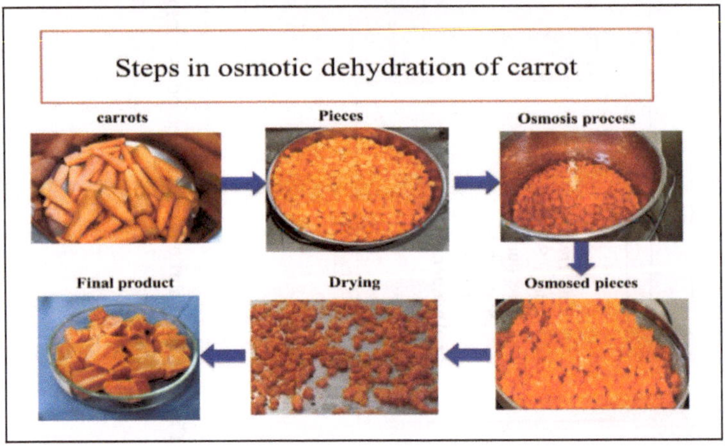

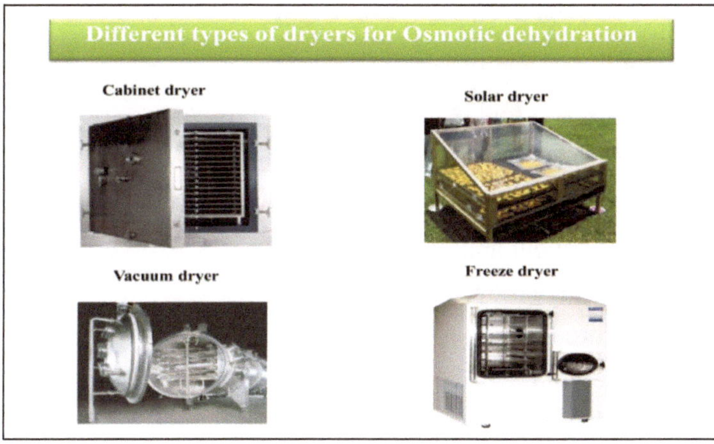

Parameters influencing osmotic dehydration

1. Properties of the Commodity
2. Pretreatments
3. Type of Osmotic Agent
4. Concentration of Osmotic Solution
5. Process Temperature
6. Agitation
7. Osmotic Solution To Food Ratio
8. Geometry of The Sample
9. Time of Treatment

Advantages of Osmotic Dehydration

- Higher retention of flavour and colour
- Enzymatic and oxidative browning is prevented
- Sweetening of the product
- Less energy consumption
- Increase the solid density
- Better textural quality
- Less expensive
- Simple equipments are required
- Storage life of product is greatly enhanced
- Marketing, handling and transport become much simpler

Limitations

- The reduction in acidity level reduces the characteristic taste of some products.
- Sugar uptake may not be desirable in certain products.

Methodology

Selection of fruits/vegetables
↓
Washing, Trimming, Peeling and Cutting
↓
Addition of sugar / syrup or salt solution
↓
Addition of citric acid (0.2 %) and KMS (1000 ppm)
↓
Keeping over night
↓
TSS of syrup increased by 10° B every day till 70° B
↓
Rinsing
↓
Drying / Dehydration
↓
Packaging
↓
Storage

Osmotic dehydrated product is known as candy, which is a sweet product prepared from fruits or vegetables by impregnating them with sugar syrup followed by draining of excessive syrup and then drying the product to a shelf stable state. The maintenance of colour, texture, taste and aroma are major problems in preparation of candy.

Vacuum Frying

What is frying?

Frying is a complex unit operation in which food is immersed in an oil bath at a temperature above the boiling point of water. This results in counter flow of water vapour (bubbles) and oil at the surface of the product. It can be considered as a method of preservation by dehydration of food.

Types of frying

In general, there are 3 types of frying methods as follow:

Pan frying

Air frying

Deep fat frying

Conventional deep fat frying

- Frying done by immersion of fruits and vegetable slices in hot oil at atmospheric pressure
- The temperature generally ranges from 130° to 190°
- Generally, 25-45% oil is absorbed by food product in deep fat frying
- It can be done in fruits and vegetables having low sugar content
- If sugar content of frying commodity is high, then browning of product will occur

Mode of heat and mass transfer during deep fat frying

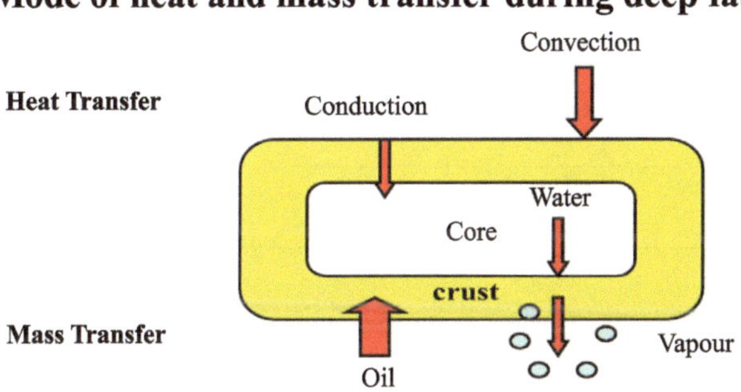

Problems associated with deep fat frying in atmospheric conditions

- High oil absorption
- Oxidation of oil and pigments
- Burning of food
- Browning of product
- Production of toxic compounds

Health problems associated with fried foods

- Increased risk of coronary heart disease
- Increased obesity risk
- Increased stroke risk
- Increased hypertension risk
- Increased diabetes risk

Vacuum frying

- Vacuum frying is a deep fat frying process that is carried out in a closed system under low pressure
- The reduced pressure makes it possible to substantially reduce the boiling point of water and thus the frying temperature.
- Most of the benefits of this technology are the result of the low temperatures employed and minimal exposure to oxygen

Vacuum frying	**Atmospheric frying**
Frying happens at low temperature *i.e.* 80-120°, so there is no burning of any ingredients	Frying happens at high temperature of about 130-190°, at this temperature burning of ingredients happens
Maximum retention of nutrients and vitamins	Nutrients and vitamins get destroyed by more than 50 %
The absorption of oil is reduced by 50% in vacuum as compared to atmospheric frying	Absorbs higher amount of oil. (generally 25 % to 45 % on weight basis)
There is authenticity of the taste	There is not originality in taste
No oxygen so no oxidation of oils and pigments	Oxygen in atmosphere is responsible for oxidation
Colour of the products remains close to original in vacuum frying	Browning occurs due to high temperature

Vacuum Frying System

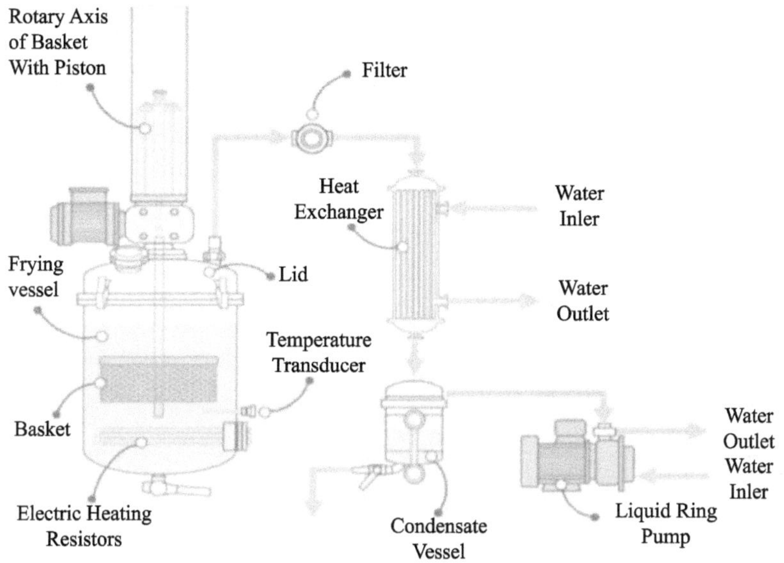

Flow chart of vacuum frying process

Fresh fruit/vegetable

↓

Peeling and slicing

↓

Pre-treatments: Pre-drying, Blanching, Anti-Browning agents, Hydrocolloids, Freezing, Osmotic-dehydration

↓

Vacuum frying

↓

Post-treatments

↓

Centrifugation

↓

Vacuum fried product

Effect of pre-treatments on quality attributes of vacuum fried products

Pre-treatments	Fruit/vegetable	Treatment description	Quality attributes			
			Oil content	**Texture**	**Nutrient**	**Color**
Blanching	Sweet potato	100 □ for 5 min	Positive	-	-	-
Hot air drying	Apple	80 □ till 64 % moisture content	Positive	N.A.	N.A.	Positive
Osmotic dehydration	Apple	Fructose 30 %	Positive	Positive	-	Negative
	Mango	Maltodextrin 40-65 % for 5 hr.	Positive	-	-	-
	Gold kiwifruit	Dextrose 55 %	-	Positive	-	Negative
Freezing	Apple	-30 □	N.A.	Positive	N.A.	N.A.
Anti-browning agent	Apple	Citric acid 5.8 %	N.A.	N.A.	N.A.	Positive
	Banana	1 % cysteine-citric acid	-	-	-	Positive
Hydrocolloids	Banana	1.5 % guar gum	Positive	Negative	N.A.	Positive

Important Process Parameters for Vacuum Frying

- Time
- Temperature
- Pressure
- Type of frying oil

Important product quality parameters

- Oil absorption
- Moisture content
- Water activity
- Breaking force
- Colour retention
- Nutrient retention
- Formation of toxic compounds

Advantages of vacuum frying

- Reduction of adverse effects on oil quality
- Preservation of natural colour and flavours
- Decreased acrylamide content
- Preservation of nutritional compounds
- Reduce fat content in product
- Some high sugar fruit/vegetable can be fried using vacuum frying

Disadvantages of vacuum frying

- Comparative high investment and operating cost
- Requires skilled operating person
- Takes little longer time

Application of vacuum frying in important horticultural produce

Product	Temperature (□)	Time (min.)	Pressure (kPa)	Centrifuged
Apple	98	4.5	6.48	no, product stood for 3 minutes
Apricot	100	72.5	2.3	no, product stood for 4 minutes
Banana	90	90	9.33	not mentioned
	89	90	2.66	140 or 280 rpm for 5 minutes
	110	20	8.0	not mentioned

Carrot	98	5	6.48	no, product stood for 3 minutes
Chinese purple yam	100	15	90	450 rpm for 5 minutes
	100-110	15	80	300 rpm for 2 minutes
Gold Kiwifruit	80	50	2.3	no, product stood for 4 minutes
	72-76	35-65	2.3	no, product stood for 4 minutes
Green Kiwifruit	105	8	6.2	750 rpm for 4 minutes
Mushrooms	90	12.5	4.25	400 rpm for 10 minutes
Potato	98	6.5	6.48	no, product stood for 3 minutes
Shallot	108	13	27.86	1000 rpm for 3 minutes
Sweet Potato	90	30	20	400 rpm, time not mentioned
	130	2.33	1.33	not mentioned

12

Food Concentration

Dev Raj[1], Shorya Raj[2], S.K. Dwivedi[3] and Avnish K. Pandey[1]

[1]ASPEE College of Horticulture, Navsari Agricultural University, Navsari Gujarat, India

[2]Food Tech Scholar, College of FPT&BE, Anand Agricultural University, Anand Gujarat, India

[3]College of Horticulture, RVSKVV Campus, Mandsaur, Madhya Pradesh

Introduction

The process of concentrating a fruit juice / pulp, vegetable juice, medicinal plant extracts or milk to a reasonably thick consistency is called concentration. Thus, concentration is a water removal process from the liquid food materials. The basic principle of concentration is boiling the liquid under partial vacuum at low temperature till desired concentration is reached. With respect to water removal, the concentration process is similar to dehydration. The food concentration is distinguished from drying on the basis of the finished product where the final product called concentrate is still liquid. In concentration process, only liquid food is used while in drying both liquid as well as solid food can be used. Fruit juice concentrates, milk concentrate are valuable semi-processed products for use in the production of value added products like fruit juice beverages and fruit juice powders. Since the water content of single strength fruit juices is very high (about 75-90%), concentration of fruit juices not only provides microbiological stability but also permits economy in packaging transportation, and distribution of the finished product due to reduction in bulk by weight and volume.

Methods of Food Concentration

The commercially feasible methods for concentration of liquid foods are evaporative concentration, freeze concentration and membrane concentration.

1. Evaporative concentration.
2. Freeze concentration
3. Membrane concentration:

1. Evaporative Concentration

Evaporation is an operation that converts a liquid into a vapour, leaving behind a concentrated solution or a high boiling liquid. Evaporative concentration is probably the oldest method of concentration and even today the most favorable and widely used method for concentration of liquid foods. Hence, it is called evaporative concentration. Evaporative concentration is the partial removal of water from liquid foods by boiling off water vapour. It increases the solids content of a food and hence preserves it by a reduction in water activity. Separation is achieved by exploiting the differences in volatility between water and solutes. Evaporation is used to pre-concentrate foods (for example fruit juice, milk and coffee) prior to drying, freezing or sterilization and hence to reduce their weight and volume. This saves energy in subsequent operations and reduces storage, transport and distribution costs. There is also greater convenience for the consumer (for example fruit drinks for dilution, concentrated soups, tomato or garlic pastes, sugar) or for the manufacturer (for example liquid pectin, fruit concentrates for use in ice cream or baked goods). Different types of evaporators are being used for drying by heating and evaporating the liquid part of the substance and these may be solar concentrator, open kettles, flash evaporators, thin film evaporators, vaccum evaporators.

Classification of Evaporators

I. On the basis of operating pressure

i) Atmospheric evaporator
ii) Vacuum evaporator

II. On the basis of effect. Effect means transfer of vapours from one evaporator to another

i) Single effect evaporator
ii) Double effect evaporator
iii) Triple effect evaporator
iv) Multiple effect evaporator.

III. On the basis of stage. Stage means flow of products from one evaporator to another

i) Single stage evaporator
ii) Multiple stage evaporator

IV. On the basis of Convection

i) Natural convection: No agitation.
ii) Forced convection: Pump is used for agitation.

V. On the basis of continuity of operation

i) Batch type evaporator

ii) Continuous type evaporator.

VI. On the basis of movement of feed

i) Forward feed evaporator

ii) Backward feed evaporator

iii) Mixed feed evaporator

Factors Affecting Rate of Evaporation

- Temperature sensitivity of materials.
- Boiling point elevation.
- Pressure and temperature.
- Solubility of the solute
- Foaming and frothing
- Fouling
- Concentration of the liquid foods.
- Rate at which heat can be transferred to the liquid,
- Quantity of heat required to evaporate each kg of water,
- Maximum allowable temperature of the liquid,
- Pressure at which the evaporation takes place,
- Changes that may occur in the foodstuff during the course of the evaporation process.

Important Practical Considerations in Evaporator

The evaporator has two principal functions (i) to exchange heat and (ii) to separate the vapour that is formed from the liquid. Therefore, the important practical considerations in evaporator are:

- Maximum allowable temperature, which may be substantially below 100°C.
- Promotion of circulation of the liquid across the heat transfer surfaces, to attain reasonably high heat transfer coefficients and to prevent any local overheating
- Viscosity of the fluid, which will often increase substantially as the concentration of the dissolved materials increases
- Tendency to foam, which makes separation of liquid and vapour difficult.

Applications of Evaporators

- It is used for pre-concentration of liquid foods prior to drying because evaporation is generally less expensive than drying. Eg. Milk concentration to make powder.
- Used for concentration of fruits juices.
- Used for concentration of salt and sugars prior to refining.

Components of An Evaporator

An evaporator being used for concentration of food consists of generally four components (Fig. 1). These are i) heat exchanger, ii) vacuum chamber, iii) vapour separator and iv) condenser.

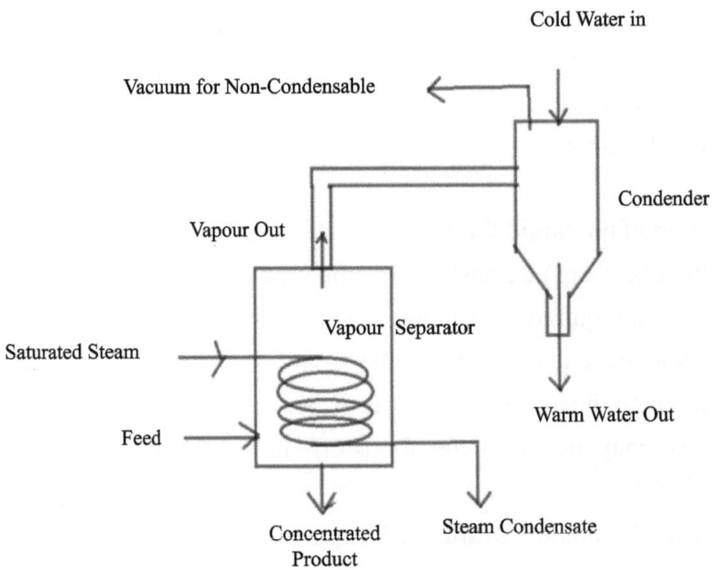

Fig. 1. Components of evaporator

Important Evaporators for Concentration

1. The Single Effect Evaporator

The evaporator which operated at single stage and has only one heat exchanger is called single effect evaporator. The typical evaporator is made up of three functional sections: the heat exchanger, the evaporating section, where the liquid boils and evaporates, and the separator in which the vapour leaves the liquid and passes off to the condenser or to other equipment. In many evaporators, all three sections are contained in a single vertical cylinder. In the centre of the cylinder there is a steam heating section, with pipes passing through it in which the evaporating liquors rise. At the top of the cylinder, there are baffles, which allow the vapours to escape but check liquid droplets that may accompany the vapours from the liquid surface. A diagram of this type of evaporator, which may be called the conventional evaporator, is given in Fig. 2. In the heat exchanger section, called a calandria in this type of evaporator, steam condenses in the outer jacket and the liquid being evaporated boils on the inside of the tubes and in the space above the upper tube plate. The resistance to heat flow is imposed by the steam and liquid film coefficients and by the material of the tube walls. The circulation of the liquid greatly affects evaporation rates, but circulation rates and patterns are very difficult to predict in any detail. However, with dissolved solids in increasing quantities as evaporation proceeds leading to increased viscosity and poorer circulation, heat transfer coefficients in practice may be much lower than this. Further, as evaporation proceeds, the remaining liquors become more concentrated and because of this the boiling temperatures rises. The rise in the temperature of boiling reduces the available temperature drop, assuming no change in the heat source. And so the total rate of heat transfer will drop accordingly. Also, with increasing solute concentration, the viscosity of the liquid will increase, often quite substantially, and this affects circulation and the heat transfer coefficients leading again to lower rates of boiling. Yet another complication is that measured, overall, heat transfer coefficients have been found to vary with the actual temperature drop, so that the design of an evaporator on theoretical grounds is inevitably subject to wide margins of uncertainty. In single effect evaporator steam economy is less than one, while in multi effect evaporator, it is greater than one. Steam economy is ratio of quantity of water evaporated per hour to steam required per hour.

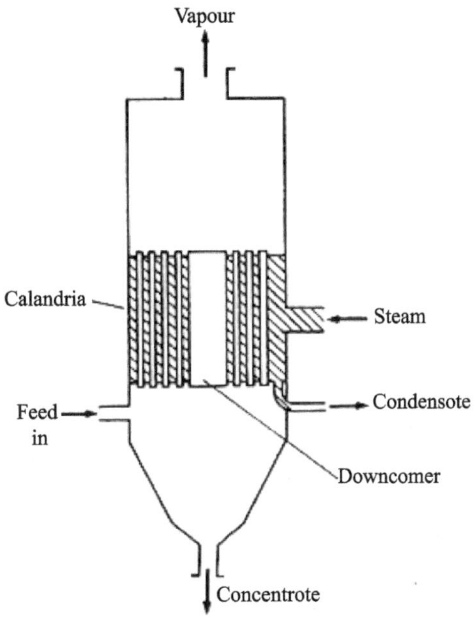

Fig. 2. Single Effect Evaporator

2. Multiple Effect Evaporators

Multi effect evaporator operates at multi stages and has more than one heat exchanger (Fig. 3 & 4). An evaporator is essentially a heat exchanger in which a liquid is boiled to give a vapour in a low pressure (Driving force) steam generator. Multiple Effect Evaporators are used for heat sensitive products where heating is required at low temperature. To make use of large amount of heat in generating vapours, same vapours can be used as heating media in next evaporation. It may be possible to make use of this, to treat an evaporator as a low pressure boiler, and to make use of the steam thus produced for further heating in another following evaporator called another effect. Second evaporation works at lower pressure than first (Double effect). Addition of another evaporator gives rise to multiple effect evaporators. Subsequent effect has lower pressure to maintain temperature difference between feed and heating media. In two effect evaporator, two evaporators are connected in such a way that the vapour line from one evaporator is connected to the steam chest of the other evaporator and is illustrated in Fig. 4.

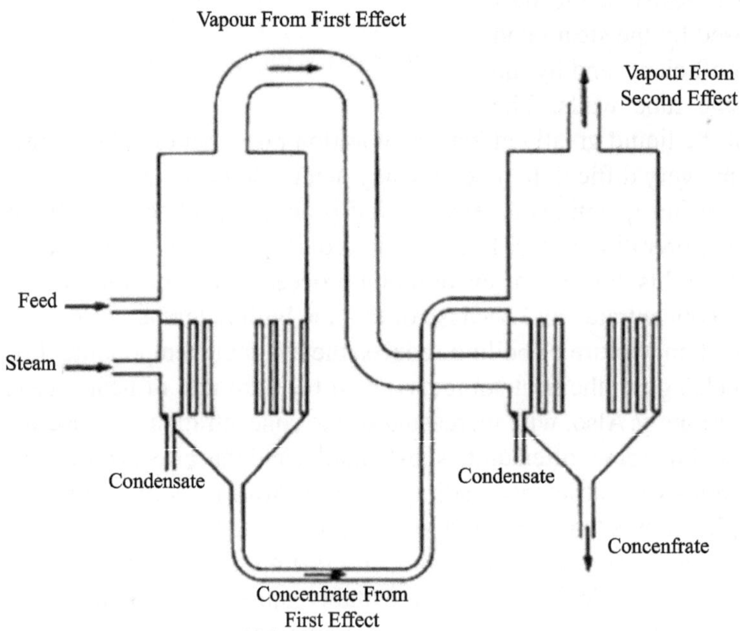

Fig. 3. Double effect evaporator

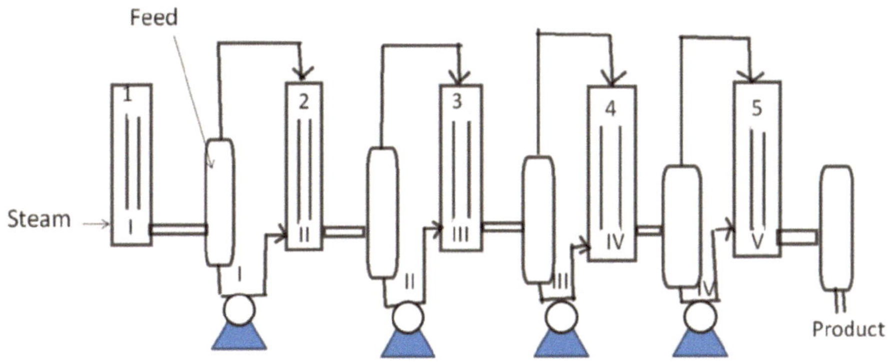

Fig. 4. Multi effect evaporators (TASTE)

In a two effect evaporator, the temperature in the steam chest is higher in the first effect than in the second effect. In order that the steam provided by the evaporation in the first effect will boil off liquid in the second effect, the boiling temperature in the second effect must be lower and so that effect must be under lower pressure. Consequently, the pressure in the second effect must be reduced below that in the first. In some cases, the first effect may be at a pressure above atmospheric; or the first effect may be at atmospheric pressure and the second and subsequent effects have therefore to be under increasingly lower pressures. Often many of the later effects are under vacuum. Under these conditions, the liquid feed progress is simplest if it passes from effect one to effect two, to effect three, and so on, as in these circumstances the feed will flow without pumping. This is called *forward feed.* It means that the most concentrated liquids will occur in the last effect. Alternatively, feed may pass in the reverse direction, starting in the last effect and proceeding to the first, but in this case the liquid has to be pumped from one effect to the next against the pressure drops. This is called *backward feed* and because the concentrated viscous liquids can be handled at the highest temperatures in the first effects it usually offers larger evaporation capacity than forward feed systems, but it may be disadvantageous from the viewpoint of product quality. Movement of feed can be of three different types. These are i) Forward feed evaporator, ii) Backward feed evaporator and iii) Mixed feed evaporator.

i) Forward feed evaporator: In Forward feed evaporator, at first the liquid fed to the first effect and is taken / flow to second and then third evaporator for heating in the same direction as vapour (Fig. 5). The liquid in forward feed system enters highest temperature evaporators (first) and exit as a concentrated liquid from the 3rd evaporator that operates at lower temperature and pressure. Liquid leaving 1st evaporator has highest temperature than 2nd evaporator.

Flow of product and steam are parallel, therefore there is a need of pump to move the solution from I, II, III because liquid flows with pressure gradient. Most concentrated product is at lowest temperature. Therefore there is no risk of damage due to heat to the product.

Advantages

a) Least expensive and simple to operate.
b) No feed pump required between effects.
c) Less risk to heat damage to more viscous foods

Disadvantage

1. Reduced heat transfer rate as the feed becomes more viscous. High viscosity (3rd effect) reduces overall heat transfer coefficient (OHTC).e.g. milk evaporators.
2. Rate of evaporation falls with each effect.
3. Best quality steam used on initial feed which is easiest to evaporate.
4. Heating of cool feed is done in the 1st effect. Therefore, less vapours are generated for each kg of the steam, thereby lowering the economy.

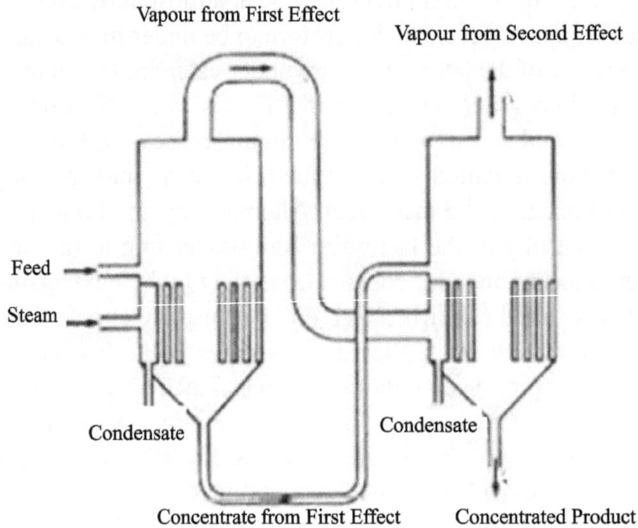

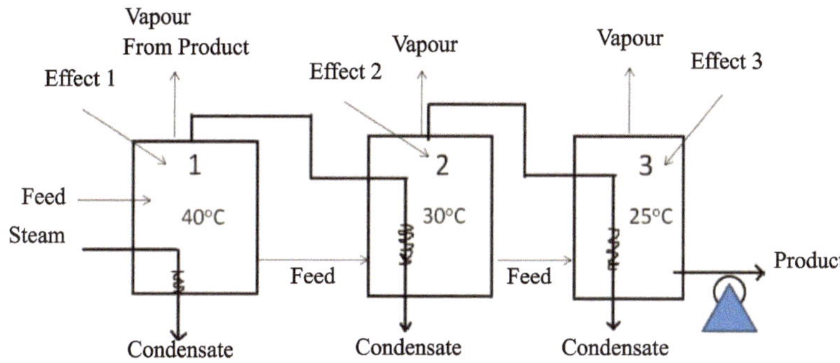

Fig. 5. Forward feed evaporator

ii) Backward feed evaporator: In this evaporator, fresh feed enters the last and the coldest effect and continues on until the concentrated product leaves the first effect (Fig 11.6). Here feed enters from the opposite direction and pumps are provided for moving the products. In backward feed system, liquid flow in opposite direction of forward feed. This system allows concentrated liquid is allowed to emit at highest temperature which is advantageous for product having high viscosity concentrated form because liquid leaving effect III is at lowest temperature than effect II.

Advantages

a) Best quality steam used on most difficult material to concentrate.

b) Better economy

c) No pump required initially.

Disadvantages

a) Inter-stage pumps are required.

b) Higher risk of heat damage of viscous foods.

c) Risk of fouling.

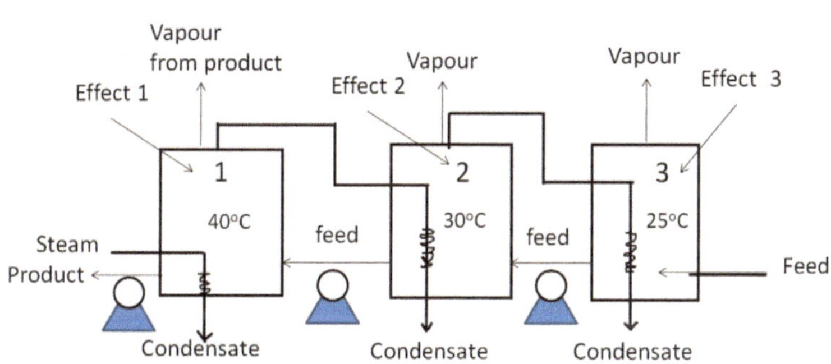

Fig. 6. Backward / Reversed feed evaporator

iii) Mixed feed evaporator: This has characteristics of both backward and forward feed. In mixed feed liquid enters II effect flow in backward feed and then pump to last effect for final concentration. This system eliminates use of pump needed in backward feed and permits final evaporation done in last effect (Fig. 7).

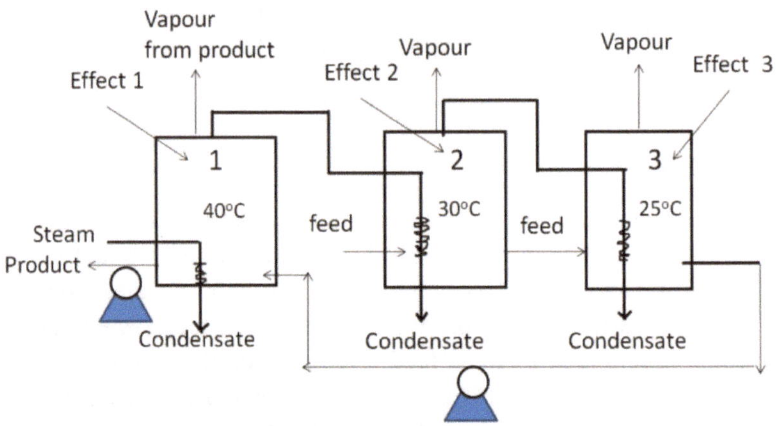

Fig. 7. Mixed feed Evaporator system

Advantages

a) Simple to operate

b) Useful for viscous foods.

c) Better economy

Disadvantages

a) More complex and expensive.

TASTE Evaporator: It is called thermally accelerated short time evaporator. These are high temperature short time (HTST) evaporator and mainly used for concentration of citrus juices. It is multi-effect evaporator and is 5 effect and 4 stages evaporator. In the figure of multiple effect evaporator; roman numbers (I, II, III and IV) shows the movement of the feed (i.e. stage) and numerical number (1, 2, 3, 4, and 5) shows the movement of vapour (effect).

Heat Transfer in Evaporators

Heat transfer in evaporators is governed by the equations for heat transfer to boiling liquids and by the convection and conduction equations. The heat must be provided from a source at a suitable temperature and this is condensing steam in most cases. The steam comes either directly from a boiler or from a previous stage of evaporation in another evaporator. Major objections to other forms of heating, such as direct firing or electric resistance heaters, arise because of the need to avoid local high temperatures and because of the high costs in the case of electricity. In some cases the temperatures of condensing steam may be too high for the product and hot water may be used. Low-pressure steam can also be used but the large volumes create design problems. Calculations on evaporators can be carried out combining mass and energy balances with the principles of heat transfer. The concept of an overall heat transfer coefficient is used in the calculation of the rate of the heat transfer (W/m^2K). The general equation for rate of heat transfer can be written as follows:

$$Q = UA\Delta T = A(Ts-Tb)$$

Where,

Q = Rate of heat transfer (W)

U = Overall heat transfer coefficient (W/m^2K)

A = Heat transfer area (m^2)

Ts = Temperature of the condensing steam (K)

Tb = Boiling point of the liquid (K).

Evaporation Equipment

- Open Pans
- Horizontal-tube Evaporators
- Vertical-tube Evaporators
- Plate Evaporators
- Long-tube Evaporators
- Forced-circulation Evaporators
- Evaporation for Heat-sensitive Liquids (TASTE)

These evaporators may be batch type or continuous type in operation. Batch Type Operation is used when small quantity has to be handled i.e. on lab scale basis. This offers greater flexibility and has low capital cost and simpler in operation. It have disadvantages of higher labour cost, increased operational cost for energy and water, less efficient use of material and energy, more floor space, lower product uniformity and quality. Continuous evaporators offer lower flexibility and capital cost is high. It has several advantages. It saves energy, space and labour. The process in under control and product obtained is uniform in quality The continuous evaporators may range from 100 kg / hour (Pilot plant) to 200,000 kg/ hy (industrial scale). First commercial evaporator was designed and used by Borden (1853).

1. **Batch type open Pans / open kettle evaporator:** The most elementary form of evaporator consists of an open pan in which the liquid is boiled. It is simple double jacketed kettle (Fig8). Heat can be supplied through a steam jacket or through coils, and scrapers or paddles may be fitted to provide agitation. The pan is fitted with a lid for vacuum operation and a stirrer to increase the rate of heat transfer to prevent food from burning. It is used in preparation of sauces, gravies, jams etc. It is also called as batch type evaporator. Such evaporators are simple and low in capital cost, but they are expensive in their running cost as heat economy is poor. It has low heat transfer coefficient under natural convection therefore agitator must be used.

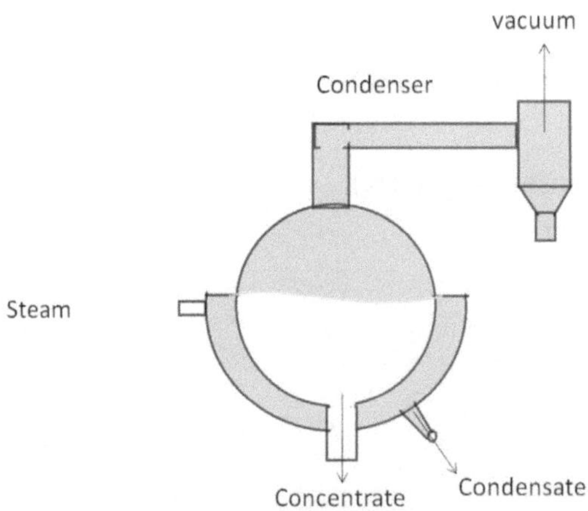

Fig. 8. Batch type pan evaporator

2. **Vacuum Evaporator:** For the evaporation of liquids that are adversely affected by high temperatures, it may be necessary to reduce the temperature of boiling by operating under reduced pressure. When the vapour pressure of the liquid reaches the pressure of its surroundings, the liquid boils. The reduced pressures required to boil the liquor at lower temperatures are obtained by mechanical or steam jet ejector vacuum pumps, combined generally with condensers for the vapours from the evaporator. Mechanical vacuum pumps are generally cheaper in running costs but more expensive in terms of capital than are steam jet ejectors. The condensed liquid can either be pumped from the system or discharged through a tall barometric column in which a static column of liquid balances the atmospheric pressure. Vacuum pumps are then left to deal with the non-condensable, which of course much less in volume but still has to be discharged to the atmosphere.

3. **Horizontal-tube Evaporators:** The horizontal-tube evaporator is a development of the open pan, in which the pan is closed in, generally in a vertical cylinder. The heating tubes are arranged in a horizontal bundle immersed in the liquid at the bottom of the cylinder. Liquid circulation is rather poor in this type of evaporator.
4. **Vertical-tube Evaporators:** By using vertical, rather than horizontal tubes, the natural circulation of the heated liquid can be made to give good heat transfer. The standard evaporator is an example of this type of evaporator (Fig. 9).

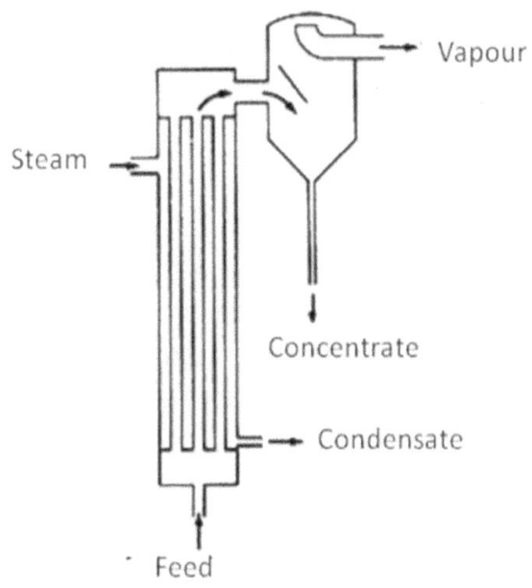

Fig. 9. Vertical-tube Evaporators

5. **Long tube Evaporators:** Tall slender vertical tubes may be used for evaporators. The tubes, which may have a length to diameter ratio of the order of 100:1, pass vertically upward inside the steam chest. The liquid may either pass down through the tubes, called a *falling-film evaporator*, or be carried up by the evaporating liquor in which case it is called a *climbing-film* evaporator. Difference between climbing film and falling film evaporator is presented in Table 1. Evaporation occurs on the walls of the tubes. Because circulation rates are high and the surface films are thin, good conditions are obtained for the concentration of heat sensitive liquids due to high heat transfer rates and short heating times.

Table 1: Difference between climbing film and falling film evaporator

Sr.No	Climbing film evaporator	Falling film evaporator
1	Fairly high degree of temperature difference is required to move the product upwards. Large temperature difference leads to scaling of tubes.	No energy is required for moving product because the flow is with gravity and as a result it leads to lower risk of scale formation in tubes.
2	Greater residence time (3-4 min) and so product quality is deteriorated	Short residence time (20-30 second) and so good quality of product is obtained
3	Foam formation in the separator due to high temperature difference.	Low temperature difference is favourable to vapour recompress hence low energy consumption.
4	Used for less viscous liquids	Can be used for more viscous liquid than as required in climbing film evaporator. Suitable for heat sensitive products like orange juice concentrate.

6. **Plate Evaporators:** The plate heat exchanger can be adapted for use as an evaporator. The spacing can be increased between the plates and appropriate passages provided so that the much larger volume of the vapours, when compared with the liquid, can be accommodated. Plate evaporators can provide good heat transfer and also ease of cleaning.

Effect on foods

- Aroma compounds that are more volatile than water are thus lost during evaporation. This reduces the sensory characteristics of most concentrates; in fruit juices this results in a loss of flavour, although in some foods the loss of unpleasant volatiles improves the product quality (for example in cocoa (Anon., 1981) and milk).
- Some volatiles can be recovered and retained in the product:
- By volatile recovery by vapour condensation and fractional distillation
- By stripping volatiles from the feed liquor with inert gas and adding them back after evaporation.

Recent Developments in Evaporative Concentration

1. *Evaporators:* The choice evaporator for concentrating a given material is important to heat sensitivity of the product is of particular importance in selecting a evaporator as it affects the quality of the concentrate. Heat damage of the product is proportional to the residence time in the evaporator heating surface. Therefore, evaporator having short residence time have been developed eg TASTE (Thermally accelerated short time evaporators) and centrifugal evaporators.

2. *Fouling:* The deposition of burn layer of organic matter in the hot surface of the evaporators is a severe problem during concentration of fruit juices particularly pulpy and cloudy juices. Rising film evaporator causes less fouling than falling film evaporator due to improper distribution of the liquid in the latter, which result in local hot spot and scaling. Agitated thin film and rotary coil evaporators also minimize the fouling.
3. Cloudy fruit juices contain variable amount so suspended particles comprising mainly of pectic substances cellulose and hemicelluloses, which increases the viscosity. Cloudy juices, centrifugal evaporators or agitated thin film evaporators are developed. Concentration of pulpy juices: pulpy fruit juices almost behave as non new twain pseudo plastic fluids and the apparent viscosity of these fluids decreases with increasing shear rate. Therefore, evaporators having proper agitation are well suited for concentration of pulpy juices. Serum concentration process is an alternate method for concentration of pulpy fruit juices. In this method the fruit juice is centrifuged to separate the solid phase (pulp) and the liquid phase (serum) is then concentrated in an evaporator before mixing with the pulp. It has been reported that the quality of higher 0Brix mango concentrate obtained by serum concentration procedure was much better than that obtained by the connectional methods.
4. *Aroma recovery:* In this process the volatile aroma compounds are transferred from fruit juices to another phase. This second phase can be gas (Evaporative gas stripping) a liquid (solvent extraction) or a solid (adsorption). In the second step of the process the separated aroma is concentrated to aroma rich solution which can be added back to the concentrated juice (Fig 10).

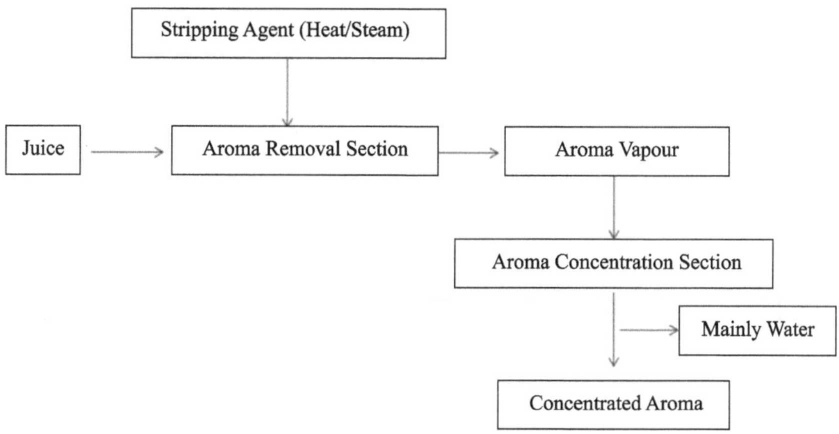

Fig. 10: Aroma recovery by evaporation distillation process

2. Freeze Concentration

Freeze concentration is a cold and selective concentration procedure, in which two distinctive steps viz., ice crystallization and ice separation from concentrate phase are involved. In the first stage, fruit juice is super cooled below its freezing point to allow water to separate as ice crystals (Fig 11). In the second stage the ice crystals are separated from the concentrated fruit juices. Frozen concentrate is a juice product from which a substantial portion of water has been removed by high vacuum and then frozen and packaged.

Freeze concentration is effective because a solute in solution has a lower melting point than pure ice, and can be separated from the water component (ice) as the temperature approaches the melting point of pure ice. To easily confirm this, remove a plastic jug of cider from the freezer, invert the jug in a suitable container, place the jug and container in a refrigerator and allow it to approach thawing temperature. Concentrated apple juice will collect in the container, leaving ice in the jug.

Advantages

Freeze concentration results in higher retention of volatile flavor constituents in the concentrate than any other concentration process, including freeze drying. Basic physical principles account for this unique feature. During freeze concentration, latent heat is removed from the product and water in the crystalline form is separated from the dissolved solids. By other concentration processes except reverse osmosis, latent heat is added to the product and water vapor is removed. Even if the volatile flavors driven off by evaporative processes are recovered and returned to the product, the heat required to evaporate the water can easily cause change in flavor and result in some loss of vitamin content. Furthermore, there is an additional cost in recovering the volatile materials and in blending them back into the finished concentrate.

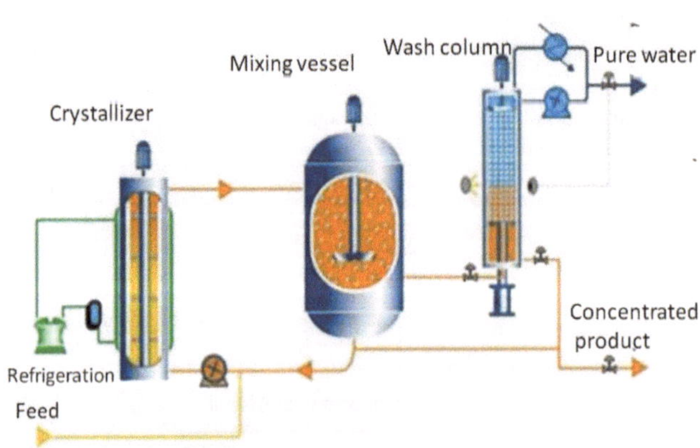

Fig. 11. Freeze concentration

Recent Developments: Most of the recent developments have been carried out for overcoming the drawbacks of single stage freeze concentration methods. The single stage freeze concentration method consists of one crystallization and one separation step. Due to this, separation of ice crystals becomes difficult as a result of increased viscosity of the end of concentrate on the other hand in the multistage freeze concentration process the ice crystals are separated out at the end of each cycle and the remaining concentrate is fed to one of the major problem in freeze concentration is the loss of soluble solids of the juice in the separated ice. This problem however has been significantly over come by the development of wash column for the efficient separation of concentrate from the ice crystals. Another problem is that the final concentration obtained is a low as 40-55% of dry matter. To reduce this juice is required to be depectinized and filtered.

3. Membrane Concentration

The use of membrane filtration on food processing applications has steadily increased in the last four decades. Some of the applications include separation, concentration, fractionation, cold pasteurization and process water treatment. The main food applications using membrane are the ones in the dairy industry, refining of crude vegetable oils, protein concentration, clarification of fruit juices and alcoholic beverages, and enzymatic reactions. Reverse osmosis (RO) and ultra-filtration (UF) are the most versatile separation processes in the food industry.

i) **Reverse osmosis:** RO is essentially a concentration process in which water is removed from low molecular weight solutes. UF is a clarification or fractionation process in which smaller solutes are transported across the membrane along with water and the membrane retains only large molecules (eg. Proteins and colloids) depending upon the pure size of the membrane.

ii) **Ultra-filtration:** Microfiltration (MF) is yet another membrane process which is mainly used for clarification purposes due to large membrane pore size. Ultra filtration is the filtration through a medium (semi permeable capillary wall) which allows small molecules (water and solute with small particle size) to pass but holds back larger ones (protein). It is used for concentration of protein only. The membrane is less tight and it restrict only macromolecules i.e. proteins and allow to pass water, sugar, vitamin etc. Ultra-filtration resists macro-molecule, while RO resist micro-molecules and allows passage of H_2O only.

Mechanism: The mechanism of separation in RO is complex and cannot be explained by molecular size only. The preferential sorption capillary flow

(PSCF) mechanism is the most logical and probable explanation, in which the mechanism takes into account both capillary flow model and solution diffusion model mechanism. In membrane separation process, a liquid is enclosed in a system confined by a semi- permeable membrane. This membrane is a barrier which allows some of its components to pass through the membrane and some cannot. The stream containing the components passing through the membrane is called permeate or filtrate and the stream containing the retained components is called retentate or concentrate.

Recent Developments

1. Among the membrane developed during last decade, Thin Film Composite (TFC) membranes of polyamide family have higher flux rate, concentration range and retention of flavour components than those with cellulose acetate (CA) membranes.
2. Fouling: Different types of membrane cleaning methods have been developed. Cleaning time of pectin fouled membrane could now be reduced to 2 min instead of 60 min by using 0.003% polyethylene oxide in 0.5% sodium hydroxide solution.
3. Low solids content achieved by conventional RO can be overcome by various techniques/processes developed.
    - Removal of suspended solids by UF prior to RO concentration allows for concentration of oranges juice to level above 42 ^{0}Birx.
    - A new process scaled fresh note process has also been developed by Septra system. By this 45-55 °Brix on commercial scale and up to 70^{0}Brix on pilot scale achieved.
    - Application of direct osmosis (DO) in fruit juice concentration is attracting interest in the recent years. Cloudy and pulpy juice should be concentrated upto 42^{0}Brix without the need for pre-filtration.

13

Innovations in Food Fermentations

Vikas Kumar

Department of Food Science and Technology, Punjab Agricultural University Ludhiana, Punjab, 141004, India

Introduction

Fermentation is playing an important role in the human life and is evident from the fact that food can be both spoiled and made by fermentation. A wide range of the fermented food products are consumed by the consumers in their daily routine. It is one for the oldest and ancient method of food preservation. It may have been a mere accident when people first experienced the taste of fermented food. Fermentation became popular with the dawn of civilization because it not only preserved food but also gave it a variety of tastes, forms, and other sensory sensations. Fermentation, along with salting, cooking, smoking, and sun drying, is one of the earliest ancient traditions developed by cultures all around the world to extend the possible storage time of foods. Slowly, people have realized the nutritional and therapeutic value of fermented foods and drinks, and this has made fermented foods even more popular. Bread, beer, wine, and cheese originated long before Christ. Although modern food technology has contributed to the present day high standard of quality and hygiene of fermented foods, the principles of the age-old processes have hardly changed. In industrialized societies, a variety of fermented foods are very popular with consumers because of their attractive flavor and their nutritional value. During food fermentations, the controlled action of selected micro-organisms is used to alter the texture of foods, preserve foods by production of acids or alcohol, or to produce subtle flavours and aromas which increase the quality and value of raw materials. After drying, fermentation is the world's oldest food preservation method.

Fermentation is described as a process in which microorganisms change the sensory and functional properties of a food to produce an end product that is desirable to the consumer. As a process, fermentation consists of transformation of simple raw materials into a range of value-added products by utilizing the phenomenon of growth of microorganisms and/or their activities on various substrates. For conduction of a smooth and effective fermentation, knowledge of microorganisms is essential. Such knowledge has only existed since 1680, when Antony Van Leeuwenhoek first demonstrated the use of a microscope and described the existence of microorganisms. Louis Pasteur, in the middle of the nineteenth century, contributed significantly to the understanding of the phenomenon of fermentation; he established the role of microbes in fermentation and also proved that there are many different kinds of fermentations. A complete detail of the history of food fermentation along with milestones is given in table 1.

Table 1. Milestones in the History of Fermented Foods

Milestone	Development/Location
ca. 10,000 B.C. to Middle Ages	Evolution of fermentation from salvaging the surplus, probably by pre-Aryans
ca. 7000 B.C.	Cheese- and bread making practiced
ca. 6000 B.C.	Winemaking in the Near East
ca. 5000 B.C.	Nutritional and health value of fermented milk and beverages described
ca. 3500 B.C.	Bread making in Egypt
ca. 1500 B.C.	Preparation of meat sausages by ancient Babylonians
2000 B.C.–1200 A.D.	Different types of fermented milks from different regions
ca. 300 B.C.	Preservation of vegetables by fermentation by the Chinese
500–1000 A.D.	Development of cereal–legume based fermented foods
1881	Published literature on koji and sake brewing
1907	Publication of book Prolongation of Life by Eli Metchnikoff describing therapeutic benefits of fermented milks
1900–1930	Application of microbiology to fermentation; use of defined cultures
1970–present	Development of products containing probiotic cultures or friendly intestinal bacteria

Source: Farnworth (2008).

Since the time of Pasteur, there have been manifold increases in the knowledge of the microbiology, biochemistry, technology, and food engineering aspects of food fermentations. At present, we have a number of fermented foods and drinks including fermented milks, cereals, fruits, vegetables, fish, meat, and many mixed products, which have emerged from very early times. Today fermented foods continue to form major sectors of the food processing industry, including baked products, alcoholic drinks, yoghurt, cheese and soy products among many others. Table 2 represents an overview of the common fermented foods along with the respective microflora.

Table 2. Common Fermented Foods

Food	Principal Ingredient	Key Microorganisms
Wine	Grapes	Yeasts
Beer	Barley	Yeasts
Cider	Apples	Yeasts
Sake	Rice	Molds
Bread	Wheat	Yeasts
Yogurt	Milk	LAB
Cheese	Milk	LAB
Buttermilk	Milk	LAB
Kefir	Milk	LAB + yeasts
Vinegar	Grapes	Yeasts followed by *Acetobacter* or *Gluconobacter*
Tempeh	Soybeans	Molds
Soy sauce	Soybeans	Molds+ LAB+ yeasts
Pickled cucumbers	Cucumbers	LAB+ yeasts
Sauerkraut	Cabbage	LAB
Pickled olives	Olives	LAB+ yeasts
Fermented sausages	Meat	LAB+ molds

Source: Rahman (2007)

Importance of Food Fermentation

- The main advantages of fermentation as a method of food processing are: the use of mild conditions of pH and temperature which maintain (and often improve) the nutritional properties and sensory characteristics of the food
- The production of foods which have flavours or textures that cannot be achieved by other methods
- Low energy consumption due to the mild operating conditions
- Relatively low capital and operating costs
- Relatively simple technologies.

Fermentation as a Preservation Method

As new preservation techniques have been developed, the importance of fermentation processes for food preservation has declined. Yet fermentation can be effective at extending the shelf life of foods and can often be carried out with relatively inexpensive, basic equipment. Therefore, it remains a very appropriate method for use in developing countries and rural communities with limited facilities. In addition, the no dependence of fermentation on the use of chemical additives to the food appeals to the "more aware" consumer market (Rahman, 2007). The chemical composition of most foods is relatively stable; therefore, generally preservation is based on eliminating microorganisms or controlling their growth and the overall composition of the microflora. To reduce or prevent microbial spoilage of food, four basic principles can be applied:

1. Minimize the level of microbial contamination onto the food, particularly from "high-risk" sources (asepsis)
2. Inhibit the growth of the contaminating microflora
3. Kill the contaminating microorganisms
4. Remove the contaminating microorganisms

Fermentations use a combination of the first three principles. Fermentations should not be expected to sterilize substandard raw products, but rather should use high-quality substrates. Microorganisms can improve their own competitiveness by changing the environment so that it becomes inhibitory or lethal to other organ- isms while stimulating their own growth and this selection is the basis for preservation by fermentation. Fermentation improves the safety of foods by decreasing the risks of pathogens and toxins achieving the infective or toxigenic level, and extends the shelf life by inhibiting the growth of spoilage agents, which cause the sensory changes that make the food unacceptable to the consumer.

Types of Food Fermentation and Microorganisms Used in Food Fermentations

The main factors that control the growth and activity of micro-organisms in food fermentations are availability of carbon and nitrogen sources, and any specific nutrients required by individual micro-organisms, substrate pH, moisture content, incubation temperature, redox potential, stage of growth of micro-organisms and presence of other competing micro-organisms (Potter & Hotchkiss, 2012). The fermentation process is mainly divided into two broad categories: submerged fermentation and solid-state fermentation. The former has been readily employed in industries for large scale production of alcohol, organic acids, enzymes, antibiotics, vitamins, and amino acids. Solid-state fermentation has been used for the production of microbial metabolites from fungi, but suffers from limitations of operation at large scales due to operational difficulties. On the basis of the by-products produced, fermentation is divided in two categories that are homofermentation and hetero fermentation. Micro-organisms that produce a single main by-product are termed *homofermentative* whereas those that produce mixed products are *heterofermentative*.

The principles groups of microbes involved in food fermentation are lactic acid bacteria (LAB), acetic acid bacteria, yeast and molds. LAB perform an essential role in the preservation and production of wholesome foods. Examples of lactic acid fermentations include (a) fermented vegetables such as sauerkraut, pickled cucumbers, radishes, carrots, and olives; (b) fermented milks such as yogurt, kefir, and cheeses; (c) fermented/leavened breads such as sourdough breads; and (d) fermented sausages.

A second group of bacteria with importance in food fermentations are the acetic acid producers i.e. acetic acid bacteria. Acetic acid is one of the oldest chemicals known; it is named after the Latin word for vinegar "acetum". The acetic acid bacteria are acid tolerant, grow well at pH levels below pH 5.0, are Gram-negative, motile rods, and are obligate aerobes. They are found in nature where ethanol is produced from the fermentation of carbohydrates by yeasts, such as in plant nectars and damaged fruits. Other good sources are alcoholic beverages like fresh cider and unpasteurized beer. In liquids, they grow as a surface film because of their demand for oxygen. The most desirable action of acetic acid bacteria is in the production of vinegar. The same reaction can also occur in wines, when oxygen is available, and here the oxidation of alcohol to acetic acid is an undesirable change, giving the wine a vinegary off-taste.

Yeasts are widely distributed in natural habitats that are nutritionally rich and high in carbohydrates, such as fruits and plant nectars. Out of the number of species, only a small number are regularly used to make alcoholic beverages. *Saccharomyces cerevisiae* is the most frequently used and many variants are available. *Saccharomyces cerevisiae* ferments glucose but does not ferment lactose or starch directly. Yeasts are used to produce ethanol, CO_2, flavor, and aroma. Other metabolic products include minor amounts of ethyl acetate, fusel alcohols (pentanol, isopentanol, and isobutanol), sulfur compounds, and leakage of amino acids and nucleotides that can all contribute to the sensory changes induced by yeasts.

The majority of fungal species have filamentous hyphae and are referred to as molds. They are grouped into four main classes based on the physiology and production methods of their spores. Molds are aerobic and have the greatest array of enzymes. Some molds are used in the food industry to produce specific enzymes such as amylases for use in bread making. They are relatively tolerant to extreme environments and are able to colonize and grow on most foods. Molds are important to the food industry, both as spoilers and preservers of foods and in particular in fermentations for flavor development. Certain molds produce antibiotics, while mycotoxin production by others is an emerging cause of concern in the food industry.

Starter Cultures

Fermented foods may be produced by the action of fermentative microorganisms naturally found on the raw materials or in the production environment. However, to improve reliability "starter cultures" are frequently used. Starter cultures may be pure or mixed cultures. The figure 1 represents the different types of the starter cultured used in food industry.

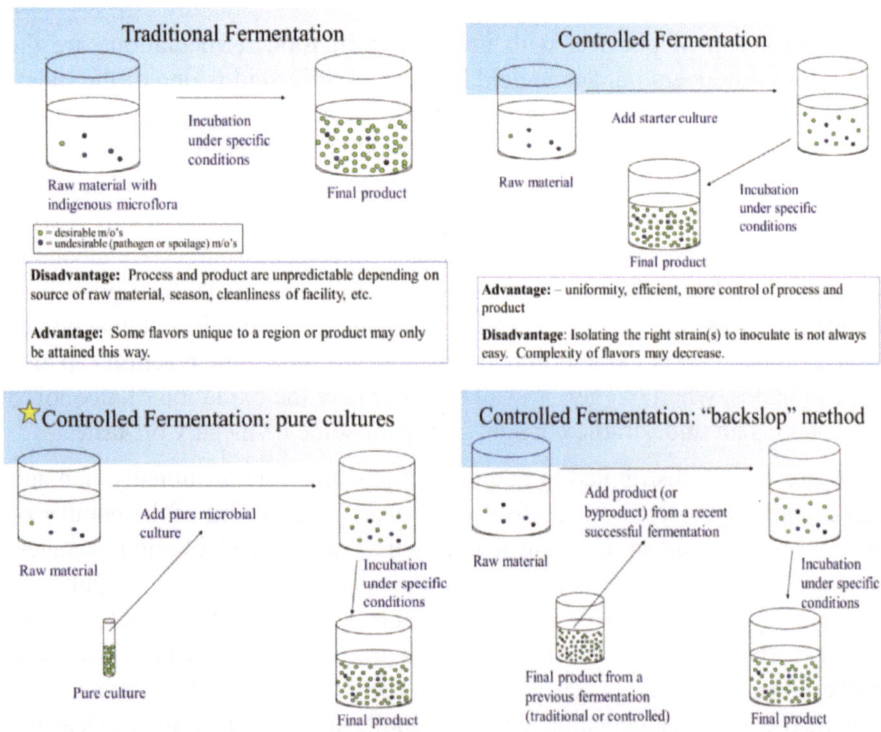

Fig. 1. Represents the different types of the starter cultured used in food industry

Fermented Products

Alcoholic Beverages

Throughout history alcoholic beverages have had a place in most cultures. They require the alcoholic fermentation of fruits or other high-sugar materials by yeasts. The alcohol content of the beverage acts as a preservative and many of these products have long shelf lives. Over the years, brewing yeasts have evolved by selection and mutations, and have been developed by genetic engineering. Major advances have been made in improving the characteristics of the fermentation strains driven by the high revenue associated with the alcoholic beverage industry.

Beer is produced by the fermentation of partially germinated cereal grains, referred to as malt, by yeasts. Beers have a final ethanol content of about 3%–8%; a huge variety of beers exists and they include ales, lagers, and stouts. Both lagers and ales can be either light or dark in appearance. Ale is produced using *Saccharomyces cerevisiae*, a top fermenter yeast, whereas lagers are produced using pure cultures of *Saccharomyces carlsbergensi*, a bottom fermenter yeast. Ales are produced using warm fermentation temperatures, 12°C–18°C and lager fermentation temperature is generally cold, 8°C–12°C. Most beer produced is of the lager variety.

Wine can be produced from any fruit juice with sufficient levels of fermentable sugars, in most cases wine is a beverage obtained by full or partial alcoholic fermentation of fresh, crushed grapes or grape juice (must), with an aging process. Wine-type grapes from cultivars of *Vitis vinifera* vines are most commonly used to produce wines. Wine making involves a series of steps. First, grape clusters are cleaned of rotten and dried berries and then separated from the stems. The grapes are subsequently crushed and pressed to release juice, the must. The remaining grape skins and seeds, called pomace, are then removed after a second press. In red wine making, the must is fermented together with the skin to extract the red pigments from the skin, which are released only during fermentation. The extraction of the red pigments is sometimes facilitated by raising the temperature to 50°C prior to fermentation of the mash, or to 30°C after the main fermentation, followed by a short additional fermentation.

The fermentations discussed above can only produce a maximum alcohol content of about 17%. Concentrations in excess of this inhibit the metabolism of the yeasts. To obtain higher alcohol concentrations, the fermented product must be subsequently distilled. Whiskey, gin, vodka, rum, and liqueurs are examples of distilled spirits. Although the process for producing most products of these types is quite similar to that for beers, the content of alcohol in the final products is considerably higher.

Vinegar and other food acids

Ethanolic fermentation of wine, cider or malt by yeast is the first of two-part fermentation in the production of vinegar. In the second stage the ethanol is oxidised by *A. aceti* to acetic acid and a number of flavour compounds. This stage is sensitive to the concentration of dissolved oxygen, and fermenters are carefully designed to ensure that an adequate supply of air is maintained (Beaman, 1967). During maturing of vinegar, reactions between residual ethanol and acetic acid form ethyl acetate, which imparts the characteristic flavour to the product.

Citric acid is widely used as an acidulant in foods. It is produced by fermenting sugar using *Aspergillus niger* in submerged culture, under conditions of substrate limitation. The production of other important food acids, including glutamic acid, gluconic acid, lactic acid, propionic acid and tartaric acid.

Lactic Acid Products

Dairy Products

There are a large number of cultured milk products produced throughout the world (for example yoghurt, cheese, Kefir, Koumiss, buttermilk, sour cream and Leben). Differences in flavour are due to differences in the concentration of lactic acid, volatile aldehydes, ketones, organic acids and diacetyl (acetyl methyl carbinol). The last is produced by fermentation of citrate in milk, and gives the characteristic 'buttery' aroma to dairy products. Changes in texture are due to lactic acid, which causes a reduction in electrical charge on the casein micelles. They coagulate at the isoelectric point to form characteristic flocs. Modifications to the starter culture, incubation conditions and subsequent processing conditions are used to control the size and texture of the coagulated protein flocs and hence produce the many different textures encountered. Preservation is achieved by chilling and increased acidity (yoghurt and cultured milks) or reduced water activity (cheese). Yogurt is a coagulated milk product obtained by lactic acid fermentation through the action of *Streptococcus thermophilus* and *Lb. delbrueckii* subsp. *bulgaricus.* Yogurt is prepared using either whole or skim milk, where the nonfat milk solids are increased to 12%–15% by concentrating the milk, or adding powdered skim milk or condensed milk. The concentrated milk is pasteurized at 82°C–93°C for 30–60 min and cooled to the starter incubation temperature of 40°C–45°C. Yogurt starter is then added at a level of around 2% by volume and incubated for 3–5 h, or until the titratable acidity of the final product reaches 0.85%–0.90% or a pH of 4.4–4.6. The yogurt is then cooled to 5°C to inhibit further acid production.

Cheese is a concentrated milk product obtained after coagulation and whey separation of milk, cream or partially skimmed milk, buttermilk, or a mixture of these products. Cheese may be consumed fresh or after ripening. Cheese is commonly made from cow, ewe, goat, or buffalo milk. The majority of cheeses are made from pasteurized milk. The use of subpasteurization heat treatment of milk or thermization is also practiced to limit heat-induced changes in milk without compromising microbiological safety. The fermentation of cottage cheese is stopped once casein precipitation has occurred and the flocs are removed along with some of the whey, but most other cheeses are pressed and allowed to ripen for several weeks or months. In the manufacture of cheddar cheese, *S. lactis* is added to milk and fermented for 30 min. Rennet is added and the culture is incubated for 1.5–2 h until the curd is firm enough to cut into small cubes. It is then heated to 38°C to shrink the curd and to expel whey. The curd is recut and drained several times, milled, salted and placed in hoops (press frames). It is pressed to remove air and excess whey, and the cheese is then ripened in a cool room for several months. Enzymes from both the micro-organisms and the cheese (including proteases, peptidases, lipase, decarboxylase and deaminases) produce compounds which give characteristic aroma and flavour.

Fermented Vegetables

A large number of vegetables are preserved by lactic acid fermentation around the world. The most important commercially fermented vegetables in the west are cabbage (sauerkraut), cucumbers, and olives. Others include carrots, cauliflower, celery, okra, onions, and peppers. Typically, these fermentations do not involve the use of starter cultures and rely on the natural flora. Brine solutions are pre- pared in the fermentation of sauerkraut, pickles, and olives. The concentration of salt in the brine ranges from 2.25% for sauerkraut to 10% for olives. The fermentation yields lactic acid as the major product. The salt extracts liquid from the vegetable, which serves as a substrate for the growth of LAB. Growth of undesirable spoilage microorganisms is restricted by the salt. Aerobic conditions should be maintained during fermentation to allow naturally occurring microorganisms to grow and produce enough lactic acid, and to prevent growth of spoilage microorganisms. For sauerkraut production, *Leuconostoc mesenteroides* grows first, producing lactic acid, acetic acid, and CO_2, followed by *Lb. brevis* and finally *Lb. plantarum* grows producing more acid and lowering the pH to below 4.0, allowing the cabbage to be preserved for long periods of time under anaerobic conditions. The LAB chiefly responsible for production of high-salt pickles are initially *Pediococcus cerevisiae* followed by the more acid-tolerant *Lb. plantarum* and *Lb. brevis*. *Leuconostoc mesenteroides* makes little contribution in the high-salt pickles but is active in the low salt pickles.

Fermented Animal Products

The primary reason for developing methods to ferment meats and fish was to extend the shelf life of these highly prized, perishable foods. Gram-positive micrococci have an important role in these fermentations. Several products became popular, including fermented sausages, fish sauces, and fish pastes. Many of the traditional fermentation methods are still used although the primary reason for their use is no longer preservation, but because the products are popular for their enhanced flavors. A variety of procedures for producing stable, fermented meat sausages have developed around the world. In general, preservation of the meat is achieved by adding salts and the generation of lactic acid by bacteria, which leads to a rapid fall in the pH. Micrococci, staphylococci, and yeasts are responsible for the development of color, taste, and flavor during the fermentation. In addition to fermentation, sausage processing may also include curing, smoking, drying, and aging to improve both the flavor and the shelf life. In addition to the major inhibitory factors of low-water activity achieved by the addition of salt, and in some cases drying and the accumulation of lactic acid, a "hurdle effect" is created by a combination of other inhibitors, contributing to the preservation of sausages.

Bread

The fermentation and baking of cereal flours alter the texture and flavour of the flour and make it palatable as a staple food. Fermentation has no preservative effect and the main function is to produce carbon dioxide to leaven and condition the dough. Yeast and other micro-organisms (e.g. *Lactobacillus* spp.) present in the dough also contribute to the flavour of the bread. Carbon dioxide is retained within the loaf when the gluten structure is set by heat above 74°C. The heat treatment and reduction in water activity preserve the bread.

Soy products

Soy sauce and similar products are made by a two-stage fermentation in which one or more fungal species are grown on a mixture of ground cereals and soy beans. Fungal proteases, α-amylases and invertase act on the soy beans to produce a substrate for the second fermentation stage. The fermenting mixture is transferred to brine and the temperature is slowly increased. Acid production by *P. soyae* lowers the pH to 5.0, and an alcoholic fermentation by *S. rouxii* takes place. Finally the temperature is gradually returned to 15°C and the characteristic flavour of soy sauce develops over a period of 6 months to 3 years. The liquid fraction is separated, clarified, pasteurised and bottled. The final product is preserved by 2.5% ethanol and 18% salt. In the production of tempeh, soy beans are soaked, deskinned, steamed for 30–120 min and fermented. Enzyme activity by *Rhizopus oligosporus* softens the beans, and mycelial growth binds the bean mass to form a solid cake. The fermentation changes the texture and flavour of soy beans but has no preservative effect. The product is either consumed within a few days or preserved by chilling

Benefits of Food Fermentation

Microorganisms *per se* can be used as food sources, but in many instances it is their effects on other food sources that are of major interest. The acceptability of a food to the consumer is based mainly on its sensory properties. The sought-after sensory properties of fermented foods are brought about by the biochemical activity of microorganisms. Fermented foods were developed simultaneously by many cultures for two main reasons: (i) to preserve harvested or slaughtered products, which were abundant at certain times and scarce at others and (ii) to improve the sensory properties of an abundant or unappealing produce.

Table 3. Potential Benefits of Fermented Foods

Increased	**Lowered**
Safety	Toxicity
Health benefits	Cooking time
Retail value	Production costs
Nutritional value	Equipment needs
Digestibility	Levels of antinutritional factors
Suitability for subsequent processing	
Sensory properties	
Ease of storage and transportation	
Shelf life	

References

Farnworth, E. R. T. (Ed.). 2008. Handbook of fermented functional foods. CRC press.

Fellows, P. J., 2009. Food processing technology: principles and practice. Elsevier.

https://www.med.umich.edu/pfans/_pdf/hetm-2017/0717-fermentedfoods.pdf

Joshi, V. K., and Ashok Pandey, eds. Biotechnology: food fermentation: microbiology, biochemistry, and technology. Vol. 1 and 2. Educational publishers & distributors, 1999.

Potter, N. N., & Hotchkiss, J. H., 2012. Food science. Springer Science & Business Media.

Rahman, M. S. (Ed.). 2007. Handbook of food preservation. CRC press.

14

Bio-colours: An Alternative to Synthetic Food Colorant

Shailendra K. Dwivedi*, Vivek Tiwari**, A.P. Dwivedi Dev Raj**** and Avnish K. Pandey***

**College of Horticulture, RVSKVV Campus, Mandsaur, M.P.*

***DIHAR, DRDO, Chandigarh*

****Indian Institute of Sugarcane Research, Lucknow, U.P.*

*****ASPEE College of Horticulture, Navsari Agricultural University Navsari, Gujarat-396 450, India*

Introduction

Color is an important characteristic of food. Based on the color of the food, first impressions are made: Is the fruit immature, ripe, or overripe? Is the toast burnt? Is the food fresh? Based on these first impressions, a judgment is made whether the food is safe to eat or not and whether it can be expected to taste good or not. Since color is closely associated with expectations, the addition of color to food is a way to fulfill these expectations. Color is added to food for one or more of the following reasons:

1. To replace color lost during processing,
2. To enhance color already present,
3. To minimize batch-to batch variations, and
4. To color otherwise uncolored food.

Food colors can be divided into four categories:

1. Natural colors,
2. Nature-identical colors,
3. Synthetic colors, and
4. Inorganic colors. Natural colors are pigments made by living organisms.

Usually, pigments made by modification of materials from living organisms, such as caramel, vegetable carbon, and Cu-chlorophyllin (vide infra), are also considered natural though they are in fact (except for carbon) not found in nature. Nature-identical colors are man-made pigments which are also found in nature. Examples are β-carotene, canthaxanthin, and riboflavin. Synthetic colors are man-made colors which are not found in nature—these are often azo-dyes. Examples of inorganic colors are titanium dioxide, gold, and silver.

Synthetic Dyes

A second important group of dyes, the direct acid azo□dyes, were introduced from 1875 and provided shades of yellow, orange and red. They did not fade, but many ran easily. Other major and minor groups of synthetic dyes followed, and before long the whole spectrum was covered. The new dyes, essentially pure chemical compounds, were more consistent and often more brilliant and cheaper than the natural dyes previously used. However, the many other qualities needed by the user were not always available, and the synthesis of improved dyes continued to be a main preoccupation of organic chemists.

Toxicity factor

Considering the toxic effects of the synthetic dyes, there has been a renewed effort to study and implement the various natural dyes in the dyestuff industry. Primarily there are three categories of natural dyes.

Firstly those are derived from plants like indigo. Second, the ones that are obtained from animal sources called Cochineal, and the remaining are those that are got from minerals (Ocher).Natural dyes can provide the much needed alternative to the complex world of chemical dyes.

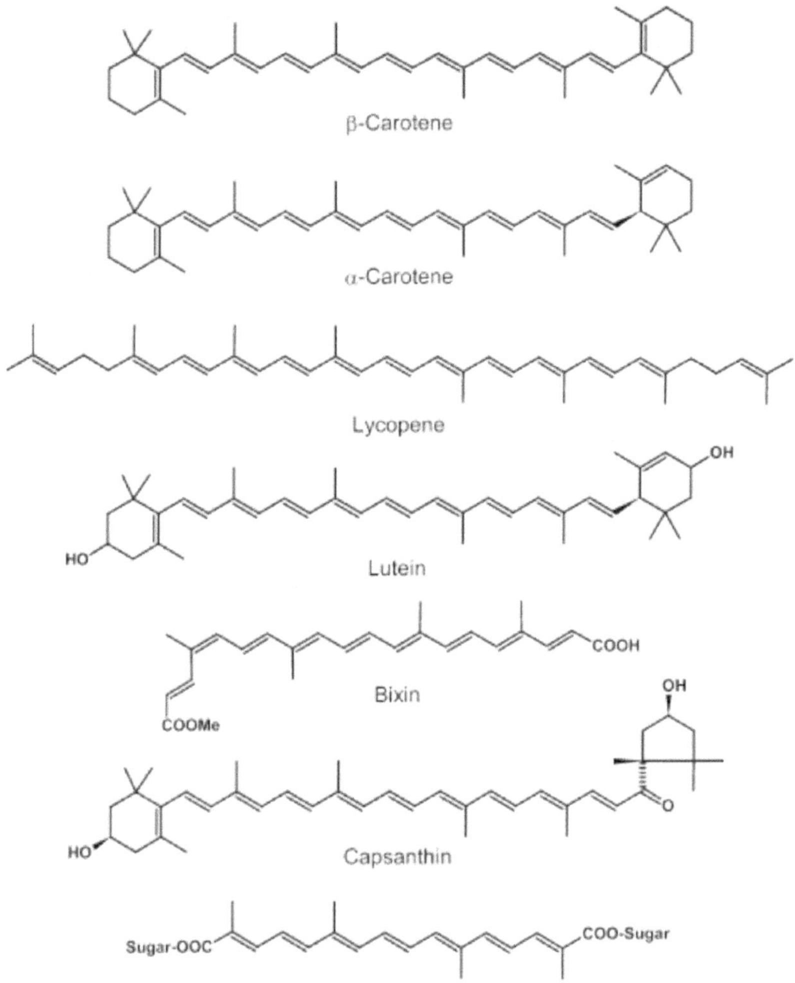

Fig. 1. Structure of some colourants

Advantages of Natural Dyes

Natural dyes cover the area of green chemistry. A great need of research is required for green dyes to replace toxic synthetic petroleum based dyes. It has been a matter of debate that synthetic dyes give a better glow and a range of colors while the natural ones are limited to dull or muddy colors only. It is not at all true, as natural dyes not only give us the feel of superior quality sensory experience but also provide a spectrum of colors. The five classic and popular natural dyestuffs are indigo, madder, cochineal, weld, and cutch. These can give rise to almost any color with the exception of a few colors like fluorescent and electric blues. Natural colors are basically plant dyes, but some of them are of animal origin as Cochineal, which is obtained from insects. These five dyes show the properties of very strong yields, resistance to fading, relatively fast colors along with easy availability.

Health and Safety Aspects of Natural Dyes

Though all natural dyes are not 100% safe they are less toxic than their synthetic counterparts. Many of the natural dyes like turmeric, annatto and saffron are permitted as food additives. Many natural dyes have pharmacological effects and possible health benefits.

1. They are obtained from renewable sources.
2. Natural dyes cause no disposal problems, as they are biodegradable.
3. Practically no or mild reactions are involved in their preparation.
4. They are unsophisticated and harmonized with nature.
5. Many natural dyes have the advantage that even though they have poor wash fastness ratings, they do not stain the adjacent fabrics in the washing process because of the non□substantive nature of the dye towards the fabric. An exception to this is turmeric, which shows substantivityfor cotton.

Use of Natural Colours in Few Countries

The natural colorants included in the United States (US) category are annatto extract, β-apo-8′-carotenal, β-carotene, beet powder, canthaxanthin, caramel color, carrot oil, cochineal extract (carmine), cottonseed flour (toasted, partially defatted, and cooked), ferrous gluconate, fruit juice, grape color extract, grape skin extract (enocyanin), paprika, paprika oleoresin, riboflavin, saffron, titanium dioxide, turmeric, turmeric oleoresin, vegetable juice. Examples of some exempt natural colourings include grape skin extract for restricted use only in beverages; marigold (Tagetes) meal can be used only for chicken food. The list of natural colourants with their specific properties is shown in Table 1.

Table 1. List of natural colourants with their specific properties

Natural food colourant	Colour hue	Source and properties
Curcumin, turmeric, turmeric oleoresin	Orange-yellow	Plant rhizome (*Curcuma longa*); strong odor and bitter, sharp taste
Riboflavin (ii) riboflavin-5-phosphate	Yellow-green	Semi-synthetic, obtained from bacterial fermentation (*Blakeslea trispora* and *Ashbya gossypi*); fluorescent lightsensitive, bitter taste

Carminic acid and carmine, Cochineal extract	Orange to red, pink to red	Female cochineal insect
Chlorophyll grass	Green to olive	Lucerne and nettle; labile to photo-oxidation
Chlorophyllins (copper complex of chlorophyll)	Bluish green	Grass, lucerne, and nettle; not claimed as natural on food labels
Caramel colours (I, II – sulfite, III – ammonia, IV – ammonia sulfite)	Brown	Food-grade carbohydrates
Carbo medicinalis (vegetable carbon)	Black	Burned plant material
Carotenes (mixtures including β-carotene), β- carotene	Yellow to orange	Palm oil or fermentation of *Blakeslea trispora* and *Ashbya gossypi*; good pH stability, oil-soluble, easily oxidized
Bixin, Norbixin Annatto extracts	Orange-red	Seeds of annatto (*Bixa orellana*)
Capsanthin, capsorubin, paprika, paprika oleoresin	Reddish (Carmine) orange	Paprika (*Capsicum annum L.*)
Lycopene	Orange red	Tomato (*Lycopersicum esculentum*)
β-Apo-8-carotenal	Orangish red	Semi-synthetic, from β-carotene
Ethyl ester of β-apo-8-carotenoic acid	Orangish red	Semi-synthetic, from β-carotene
Canthaxanthin	Orange and pink	Salmon, shrimp; semi-synthetic from β-carotene
Lutein	Orange and yellow	Aztec marigold *(Tagetes sp.)* petals
Betanin, Dehydrated beets	Pink to red	Red beet root; heat, light and oxygen-sensitive
Anthocyanins, grape color extract, grape skin extract	Pink-red to mauve-blue	Elderberries, black grape skins, black carrots, red cabbages; pH-dependent, heat-and oxidation-sensitive
Calcium carbonate	White	Mineral rocks; surface dye
Titanium dioxide	White	Mineral rocks; surface dye
Iron oxides and hydroxides, synthetic	Yellow, red, black	Mineral rocks; surface dye
Aluminium	White	Mineral rocks; surface dye

Silver	White	Surface dye
Gold	Yellow	Surface dye
Rubin BK	Red	Surface dye
Paprika	Red	Paprika
Saffron	Yellow-orange	Saffron

Purpose of Food Coloring

1. Enhancing naturally occurring colors.
2. Protecting flavors and vitamins from damage by light.
3. Decorative or artistic purposes such as cake icing.
4. Masking natural variations in color.
5. Offsetting color loss due to light, air, extremes of temperature, moisture, and storage conditions.
6. Providing identity to foods.
7. Providing varieties of wholesome and Nutritious food that meets consumer's demand.
8. Improves taste
9. Added to food or drink to change its color
10. Improves and maintains nutritional value of food

Types of Natural Colourants

Based on Their Origin

Plant / Vegetable Origin

These are the colourants derived from leaves, roots, bark, fruits, flowers, trunk of plants and vegetables either in dried or wet form produces colour. In India there are about 500 species of plants and vegetables which are found suitable to give colourants. These colourants have applications in food colouring, medicines, paper colouring etc. Examples are yellow from turmeric, red from madder roots, blue from water lily etc.

Insect / Animal Origin

These are the colourants obtained from secretion and dried bodies of insects or animals. Deep violet colour obtained from the secretion of sea mollusc, Murex is oldest known animal dye and is very expensive. Examples are lac insect secretion, urine of cow, shellfish etc.

Mineral Origin

These are the colourants obtained from earth or minerals. Oxides and hydrated oxides of manganese and iron, orchers, titanium dioxide, umbers are some examples of mineral origin natural colourants.

Microbial/Fungal Origin

These are the colourants obtained from bacteria, algae, fungi, and yeast upon micro-biological action. Colourants produced from these have anticancer and antioxidants properties. These are used in sauces, baby foods, milk products, energy drinks and many more. Factors such as temperature, pH, type of fermentation, moisture content etc. affects microbial pigment production. Examples are brown from *bacillus*, red from *dunaliellasalina*, yellow from ashbyagossypi etc.

Based on Their Chemical Constituents

Indigoid dyes

Most important dye of this class is indigo and is also the primary source of blue colour. Indigo obtained from Indigofera species is insoluble in water. For water soluble form it must be reduced to leuco form by reduction process. It is used for dyeing textiles. After dyeing, it is oxidized to blue indigotin structure which has excellent colour fastness properties. This natural indigo is similar in structure with synthetic indigo.

Anthraquinone

Red colour dyes fall into this category. Alizarin, morinda, lac, indian madder and cochineal are some examples of this dye.

Naphthoquinone

These dyes give red, orange or reddish brown shades. Walnut shell, henna etc. fall under this category of dyes.

Flavonoids

Most of yellow coloured dyes fall under this category of dyes. Luteolin, rutin, quercitrin are some examples of this dye.

Dihydropyrans

These dyes have structure like flavonoids but are substituted dihydropyrans.

Anthocyanidins

Carajurin fall under this class of dye. It is obtained from Bignonia chica.

Carotenoids

Bixin and nor bixin of annatto seeds and crocin of saffron stigma are major dyes of this class. Colour of this dye is due to presence of double bond.

Tannin based

Polyphenolic compounds of tannins fall under this class. Tannin requires mordant for textile dyeing and also changes colour with different mordant. Babool and cutch are examples of this dye.

Classification by Source

The classification of naturally derived colorants can become very complex because of the wide variety of innate properties of the coloring substances. They can be derived from a variety of sources in nature, and, therefore, natural colorants also exhibit awide variety of chemical compositions that affect properties, solubilities, and stabilities differently. These properties impact their potential applications in certain matrices and also play a role in regulations of the different pigments.

Sources of Natural Colorants

Pigments are found widespread in nature from the simplest procaryotic organisms like cyanobacteria, and throughout the kingdoms of fungi, plants, and animals. Most natural food colorants come from the division Magnoliophyta (flowering plants) of the plant kingdom. However, natural colorants from other sources such as scale insects (cochineal and lac), fungi (*Blakeslea trispora* and *Monascus spp.*), and cyanobacteria (*Arthrospira spp.*) are used today for coloring food.

Some major sources of natural colourants which are commonly used for production of natural colourants are summerised in the table:

S.No	Common name of the plant	Botanical name	Part used	Color obtained
1.	Siam weeds	Eupatorium odoratum	Whole plant	Yellow
2.	Goat weed	Ageratum conyzoides	Whole plant	Yellow
3.	Jack fruit tree	Atrocarpus heterophyllus	Bark	Yellow
4.	Gulmohar	Delonix regia	Flower	Olive green
5.	Teak	Tectona grandis	Leaves	Yellow
6.	Babool	Acacia nilotica	Leaves, bark	Yellow/ brown
7.	Water lilly	Nymphaea alba	Rhizomes	Blue
8.	Dahlia	Dahlia variabilis	Flowers	Orange
9.	Amla	Emblica officinalis	Bark, fruit	Grey

10.	Indian Jujube Bar	Ziziphus mauritiana	Leaf	Pink
11.	Drumstick	Moringa pterygosperma	Leaf	Yellow
12.	Sausage tree	Kigelia pinnata	Petals, heartwood, bark	Yellow/ Pink
13.	African tulip tree	Spathodea companulata	Flower	Yellow/ Orange
14.	Tamarind	Tamarindus indica	Leaves, seeds	Yellow/ brown
15.	Golden dock	Rumex maritmus	Seeds	Brown
16.	Eucalyptus	Eucalyptus camaldulensis	Bark	Yellow/ brown
17.	Red sandalwood	Pterocarpus santalinus	Wood	Red

Plant

Plant-derived pigments are the result of biochemical pathways within the organism that result in a variety of organic compounds with unique physicochemical properties. These colored compounds are abundant in nature and play important roles in photosynthetic pathways, attract pollinators, and provide protection from predators and solar energy. Plant pigments include a variety of chemical classes, such as porphyrins, carotenoids, anthocyanins, and betalains, which are discussed below, that can selectively absorb certain wavelengths of light while reflecting others.

Animal

Similarly, different animals produce a variety of chemical compounds that result in unique colorations across the animal kingdom. These pigments also serve a variety of purposes, including transportation of blood oxygen, protection from predators or UV radiation, mating, and more. As previously mentioned, many animal colors are the result of physical nanoscale structuring causing selective reflectance, diffraction, or scattering of certain light waves, and by pigments such as pterins, purines, anthraquinones, melanins, and others.

Microbial

Pigment production by various microbial organisms has been used to help identify certain species. Bacterial and fungal organisms produce a variety of different types of pigments, such as carotenoids and *Monascus* pigments, resulting in color expression of almost all hues. The use of microorganisms to produce colorants is commercially and economically promising because of the ability to control growth conditions and to ensure renewability.

Mineral

Minerals are generally described as elements or chemical compounds that are crystalline and formed by geological processes. They have long historical use as colorants in foods, cosmetics, and art. Minerals present a variety of hues that depend on their chemical composition and/or physical structuring. Many minerals contain metallic cations, and metallic cations containing *d* orbital electrons often absorb and reflect visible light. As an example, viridian, a chromium oxide mineral, is a green pigment that has been used in painting but also in cosmetics as a colorant that is exempt from certification.

Extraction Methods

As natural dye-bearing materials contain only a small percentage of coloring matter or dye along with a number of other plant and animal constituents such as water-insoluble fibers, carbohydrates, protein, chlorophyll, and tannins, among others, extraction is an essential step not only for preparing purified natural dyes but is also required to be carried out by users of crude dye-bearing materials. As natural coloring materials are not a single chemical entity and the plant matrix also contains a variety of nondye plant constituents, extraction of natural dyes is a complex process. The nature and solubility characteristics of the coloring materials need to be ascertained before employing an extraction process. The different methods for extraction of coloring materials are:

- Aqueous extraction
- Alkali or acid extraction
- Microwave and ultrasonic assisted extraction
- Fermentation
- Enzymatic extraction
- Solvent extraction
- Super critical fluid extraction.

Drying Techniques

Coloring matter extract obtained from natural sources is mostly in aqueous media. If it is to be immediately used for dyeing as in many small-scale or cottage-dyeing establishments, it can be used as such after adjustment of concentration according to the shade requirements. However, if it is to be used at a later date for dyeing or in dye extract producing units, it has to be converted into either powder form or concentrated solid rich form for long term storage and ease of transport. This also ensures uniformity of shade for the entire batch of dye powder or concentrate produced and natural dyes thus converted into powder form or liquid concentrates can be used like synthetic dyes by the industries. The following techniques are generally used for converting natural dye extracts into powder form or concentrates.

- Spray drying
- Drying under vacuum
- Freeze drying.

Spray drying is the most commonly used technique for converting natural dye extracts into powder form as it is simple and cost effective. The plant extract is sprayed into the spray chamber as fine droplets through an atomizer or spray nozzle. These droplets come into contact with hot air flowing into the chamber which removes the solvent; the resulting dry particles aggregate and fall to the bottom of the chamber where this powder can be collected. Natural dyes in readyto- use dry powder form are mostly produced by this method. Dye molecules should be sufficiently stable to heat to follow this method of drying as dry powder is exposed to dry heat from the hot air. Also many fine dye particles that are too small to settle at the chamber bottom are carried away by the hot air and are lost resulting in lower recovery. Also higher dye content is needed to get larger size particles and lower losses that may be achieved by preconcentration of the extract or addition of inert compounds such as lactose and so on.

Extracts can also be concentrated under vacuum by use of a rotary evaporator or tray dryers and the concentrated dye can be purified further by the use of various solvents. Another drying technique, lyophilization or freeze-drying can also be used for preparation of natural dye powders. In this process, the natural dye extract is subjected to freezing and water is removed from the frozen extract through sublimation by reducing the pressure. The equipment is costly and operating cost is also higher but heat-sensitive dyes can also be converted into dry powders as the process takes place at low freezing temperatures. Any one of the above techniques can be used for producing natural dye powder depending upon the dye extract properties and costs. Very pure dye extracts are now available for use in many countries especially the United States, but these are very costly and are mainly used by hobby groups for their uniqueness.

Field of Application

Bakery industries: Use in the manufacturing of Biscuits, Cookies, Cakes Pastries, Cake mixes etc.

Beverages Industries

Catering to the needs of beverage industry for use in the manufacturing of cold drinks, Juice, Dry Mixes etc. Cold drinks look & taste are changing due to adding these colors.

Confectionery Industries

Used in the manufacturing of confectionery items such as Jellies, Chewing Gum, Cream / Paste, Gums & Chews etc.

Cosmetics & Toiletries Industries

Used in the manufacturing of lipsticks, nail polishes etc.

Dairy & Ice cream Industries

These colors are used in Dairy and Ice cream manufacturer. These colors are being widely used in Milkshakes, Processed Cheese and Water Ices etc. These colors make the food items more delightful. These colors are prepared under the hygienic conditions, keeping into mind PH factor.

Meat & Savories Industries

These colors are highly used in Fresh comminuted Meat, Seasonings etc. These colors make food colors appetites more. It is very important for food colors to be heat resistance These colors make the foods items more delightful. These colors are food additives and hygienic in use.

Pharmaceutical Industries

These colors are most commonly used as coloring agents in food and pharmaceutical industries as they are highly adaptable and versatile. The pharmaceutical colors are available in different types such as Injectables, Tablets, Mouth wash, Gels etc. These colors are ideal for coloring products including fats, oils or items lacking sufficient moisture to dissolve dyes.

Seafood Industries

Colors are used in Fish coating, Smoked fish. These colors are most commonly used in food agents. These colors are harmless and accurately fresh.

Pet Food Industries

Used in the manufacturing of pet foods.

Advantages of Natural Colourants

Following are the advantages of using natural colourants:

1. Eco-friendly
2. Less pollution generation
3. Non-toxic
4. Non-allergic
5. Non-carcinogenetic
6. Easy to prepare
7. Biodegradable
8. Easy to dispose

Disadvantages of Natural Colourants

Following are the disadvantages of using natural colourants:

1. Fugitive
2. Proper storage conditions required
3. Shades vary from batch to batch
4. Have no standard method for preparation
5. Not easily available
6. Complex process of dyeing

Future Prospects

Development of "new" food colorants can be divided into two categories: those covered by existing legislation and those that are not. Colorants covered by existing legislation are not really new, but encompass subjects such as new sources of known pigments (e.g., a newly discovered fruit rich in carotenoids), new ways of formulating existing pigments (e.g., blue anthocyanin), and improvement of existing sources (e.g., by breeding). Breeding and selection of high-yield strains will most likely only lead to minor improvements. Major improvements could be achieved by genetic manipulation, but this is not allowed in the EU and the naturalness of the colorant would be (partly) lost. Production of colors by fermentation has a number of advantages: cheaper production, possibly easier extraction, higher yields, no lack of raw materials, and no seasonal variations. Fermented colors are already used today: *D. salina*, *B. trispora*, spirulina, and monascus. It is not unlikely that new, fermented colors such as lycopene from *B. trispora* will become allowed in the near future. A giant leap forward in color production could be achieved by combining genetic manipulation and fermentation. Microorganisms could be made to produce colorants in high yield by inserting genes coding for the colorant—even colorants not naturally produced by microorganisms (e.g., turmeric) could be made in this way. Colorants made in this way will probably face some hurdles, at least in Europe where genetically modified food is generally viewed with a large degree of skepticism by the consumers. Finally, new colorants would have to be approved by the authorities, which is very costly because of the various toxicological studies needed to confirm the safety of a new food additive. The above gives a fairly compressed overview of the most important natural food colorants. The reader is referred to a number of excellent textbooks for more detailed information on plant pigments and food colorants.

15

Innovations in Food Packaging

Vigya Mishra

Department of Post Harvest Technology, College of Horticulture, Banda University of Agriculture & Technology, Banda, Uttar Pradesh, India

Introduction

Packaging is one of the most important processes to maintain the quality of food products for storage, transportation and end-use. The basic functions of packaging are considered as protection, containment, information and convenience. Food packaging was initially considered to increase the marketable life of the food in a container by protecting it from the influence of the environmental factors. Then it was expected to be informative, easy-to-handle and disposal-friendly with increased functionality. Current innovations of food packaging technology are geared to ensure one/ all of these properties along with better protection, more efficient quality preservation and enhanced safety of the packaged food. Innovative packaging technologiesalso prevent the rate of quality deterioration and facilitate easy handling of the produce during distribution and marketing. During distribution, the quality of the food product can deteriorate biologically and chemically as well as physically. Innovative packaging technologies are the most important and challenging technologies to overcome these limitations.These technologies require specific knowledge and appropriate training and understanding for rationally selecting the most suitable packaging for each product and intended use. Innovative packaging of foods includes active packaging, smart packaging, on-the-go packaging, intelligent packaging etc. which have been briefly discussed in this manuscript.

1. Bio-Based Packaging

The worldwide interest in bio-based polymers has accelerated in recent yearsin order to find non-fossil fuel based polymers which are biodegradable and does not cause pollution.The term "biodegradable" is used to describe those materials which can be degraded by the enzymatic action of living organisms such as bacteria, yeasts and fungi. The ultimate end products of degradation process are CO_2, H_2O and biomass under aerobic conditions while hydrocarbons, methane and biomass under anaerobic conditions. Bio-based polymers are manufactured from polysaccharides (chitosan, carrageenan, starch, alginate etc.), proteins (caseinates, whey protein isolates, soy protein, wheat gluten, corn zein etc.) and lipids.

2. Nano based Packaging

Nanomaterials are increasingly being used in the food packaging industry due to the range of advanced functional properties they can bring to packaging materials. Nanotechnology enabled food packaging can generally be divided into three main categories :*Improved packaging*(whereby nanomaterials are mixed into the polymer matrix to improve the gas barrier properties, as well as temperature and humidity resistance of the packaging);*Active packaging* (illustrated by the use of nanomaterials to interact directly with the food or the environment to allow better protection of the product); and*Intelligent/smart packaging*(designed for sensing biochemical or microbial changes in the food).

Applications in food packaging

A. **Nano Encapsulation:** Nano-encapsulation has been tested for the safe and controlled discharge of favorable live probiotic cells to boost healthy gut function. The sustainability of probiotic organism such as *Lactobacillus acidophilus*, *Lactobacillus casei*, *Lactobacillus rhamnosus*, and *Bifidobacterium*spp. which are present in the freeze-dried yogurt can be amended through nano-encapsulation with the help of calcium alginate.

B. **Nano Composite:** Nanocomposites are polymer matrices reinforced with nanomaterials such as nanoparticles. These materials have at least one dimension on the nanometer scale (10−9m) and are composed of a variety of materials including metal ions (such as silver, copper, and gold) and metal oxides (TiO_2, MgO). The assimilation of different nanoparticles such as SiO_2, clay, TiO_2, $KMnO_4$, nano cellulose, nano fibrillated cellulose and carbon nanotubes are capable of amending the packaging material in terms of mechanical and barrier properties.

C. **Nano Emulsions:** The operations such as the targeted delivery of lipophilic products for example nutraceuticals, medicines, flavor, antioxidants and antimicrobial agents are accomplished by these carriers named nano-emulsions.

D. **Edible Nano Coatings and Nano Coating Materials:** Films made from edible comprises of water-soluble polysaccharides (hydrocollids) and lipids. The most commonly used polysaccharides are cellulose derived such as alginates, pectins, starches, chitosan and other polysaccharides. Polysaccharide gives variety of properties like rigidity, firmness, thickening quality, viscosity, adhesive and gel-forming capability. Different types of lipid complexes like animal and vegetable fats have been employed for synthesizing the edible films and coatings.

E. **Nanosensors:** Packaging equipped with nano-sensors is also designed to track either the internal or external conditions of food products, pellets and containers, throughout the supply chain. For example, Nanosensors in plastic packaging can detect gases given off by food when it spoils and the packaging itself changes color to alert the consumer. These films are packed with "silicate nanoparticles" to reduce the flow of oxygen into the package and the leaking of moisture out of the package to stay food fresh.

3. Shrink Packaging

Shrink Wrapping is a packaging technique which greatly reduces the moisture loss and maintains freshness of fresh produce for longer duration. Individual shrink packaging may be considered as modified atmosphere packaging (MAP) for individual fruit. It involves sealing of fruit in a flexible film followed by heat shrinking to confine the shape of fruit or container in which fruit is kept. Generally 10-20% reduction in transpiration rate is possible by shrink packaging even under ambient conditions.

5. Active Packaging

The term "active packaging" was first applied by Labuza (1987), and may be defined as packaging which performs some desired function other than merely providing a barrier to the external environment. Active packaging is normally designed to address one property or requirement of the food or beverage. The property normally chosen is that most critical as the first limiting factor of quality or shelf life. Many new "extra" functions have been introduced in active packaging technologies which includeoxygen scavenging, carbon dioxide-absorbing, moisture-scavenging (desiccation), antimicrobial activity, atmosphere control, edibility, biodegradability etc. Active food packaging performs beyond itsconventional protective barrier property. The new active packaging systems increasesecurity, safety, protection, convenience, and information delivery. Active packagingsystems extend the shelf life of food products by maintaining their quality longer,increase their safety by securing foods from pathogens and bioterrorismand enhancethe convenience of food processing, distribution, retailing and consumption.

Active packaging systems for processed foods and beverages

Oxygen-scavenging systems have been commercialized in the form ofa sachet that removes oxygen. An oxygen-free environment can prevent foodoxidation and rancidity and the growth of aerobic microbes.

Carbon dioxidescavenging/ release systems can prevent packages from inflating due to the carbondioxide formed after the packaging process – for example, packaged coffee beans may

produce carbon dioxide during storage as a result of non-enzymatic browning reactions. Similarly fermented products such as kimchi (lactic acid fermented vegetables), pickles,sauces, and some dairy products can produce carbon dioxide after the packagingprocess. Carbon dioxide-scavenging systems are also quite useful for the productsthat require fermentation and aging processes after they have been packed.

Moisturescavenging systems Recently, the sachets have contained humectants as well as desiccants to control the humidity insidethe package more specifically. Moisture-scavenging systems that are based on desiccation are evolving to control the moisture by maintaining a specific relative humidity inside the package by absorbing or releasing the moisture.

Antimicrobial packaging systems. Antimicrobial packaging is a system that can kill or inhibit the growth of micro-organisms and thus extend the shelf life of perishable products and enhance the safety of packaged products. Edibleantimicrobial agents can be incorporated into food ingredients, while antimicrobialresources can be interleaved in the in-package headspace in the form of sachets, films,sheets or any in-package supplements, to generate antimicrobial atmospheres.Organic acids, bacteriocins, enzymes, bacteriophages, essential oils etc are used as antimicrobial agents in the packages. Besides, gaseous agents like carbon dioxide for modified atmosphere packaging, sulfur dioxide for berries and ethanol vapor for confections to inhibit the growth of microorganisms.

Intelligent Packaging

Intelligent packaging has been categorized both as a part of active packaging and as a separate entity, depending on different viewpoints. The term is used to represent class of packaging materials that can monitor the condition of either the packaged food or the food environment inside the package which includes temperature, pH, etc., and provides this information to the user. It can be interpreted as an extended function of traditional packaging which is responsible for communication with the consumer depending on its ability to sense and record the changes taking place in the food or its environment.

An added advantage of intelligent packaging is its contribution to the improvement of Hazard Analysis and Critical Control Points' (HACCP) and Quality Analysis and Critical Control Points' (QACCP) systems and hence onsite detection of unsafe food products, identification of potential health hazards and establishing strategies to abate their occurrence, ultimately leading to improved food quality. The intelligent systems can be classified into three categories; i) sensors, ii) indicators and iii) radiofrequency identification (RFID) systems.

Table 1. Innovations in intelligent packaging for foods

Sl. No.	Type of intelligent packaging system		Description
1.	Sensors	Biosensors	Food Sentinel System® and ToxinGuard® are commercial biosensor based packaging materials available in the market
		Gas sensors	Gas sensors are employed for the detection of gaseous analytes like oxygen, water vapor, carbon dioxide, ethylene, etc. inside the package. The use of pH sensitive dyes like methyl red and curcumin for the detection of basic volatile amine released from rotten meat and fish have been reported.
		Chemical sensors	These can adsorb a particular chemical on the surface and detect its presence, composition, activity or concentration have been employed as chemical sensors. Carbon based nanomaterials like graphene, carbon nanotubes and carbon nanofibers are used as chemical sensors
		Electronic Nose	Instruments have been designed to identify and classify the mixture of aromas in an odor on a repeatable basis – a function similar to that of the mammalian olfactory system. The instrument is composed of an array of sensors, either chemical sensors or biosensors, which show partial specificity to each kind of odor. The statistical methods are used to recognize simple and complex odor and produce a unique response towards each one. Successful testing of electronic nose system has been carried out in response to the odor released by fresh yellow fin tuna, vacuum packed beef, fruits and vegetables, and broiler chicken.

2.	Indicators	Freshness Indicators	These indicators provide information about the product quality by determining the chemical changes resulting from the microbial growth within the product. The microbial growth metabolites react with the indicators integrated inside the food package to give visual information regarding the product quality. E.g. FreshTag®
		Time temperature indicators (TTIs)	TTIs are labels that provide visual indication of the temperature abuse of temperature dependent perishable products, like frozen foods, during distribution and storage from the point of production to the point of consumption. There are three basic types of TTI available in market: critical temperature indicators, partial history indicators and full history indicators. E.g. VITSAB®,3M Monitor Mark®, Freshness Check®, Lifelines Freshness Monitor®, FreshCode (Varcode Ltd.) and Tempix (Tempix AB)
		Integrity Indicators	Integrity indicators function to ensure their integrity. Visual oxygen indicators are composed of redox sensitive dyes which change color with change in oxygen concentration in MAP foods. E.g. Ageless Eye®,
3.	RFID	Radiofrequency identification (RFID) systems	RFID is a tag or reader based automatic identifications system used for item identification and data accumulation without human intervention. RFID tags have some identification number stored in their databases and are able to accordingly act upon it by retrieving the information about that number from the database. RFID tags for monitoring temperature, relative humidity, pressure, pH, and light exposure of the products are already available in the market which aid in enhancing food quality and safety. These are more advanced, reliable and efficient than the conventional barcode tags for food traceability.

6. Edible films and coatings

Fig. Intelligent packaging of fresh produce

The use of edible films and coatings is an application of active food packaging, since the edibility and biodegradability of the films are extra functions that are not presentin conventional packaging systems. Edible films and coatings are useful materials produced mainly from edible biopolymers and food-grade additives. Mostbiopolymers are naturally existing polymers, including proteins, polysaccharides (carbohydrates and gums), and lipids. Plasticizers and other additives are included with the film-forming biopolymers in order to modify film physical properties or to create extra functionalities.The application of edible films and coatings is an easy way to improve the physical strength of the food products, reduce particle clustering, and enhance the visual and tactile features of food product surfaces. The most common functions of edible films and coatings are that they are barriers against oils, gas or vapours, and that they are carriers of active substances such as antioxidants, antimicrobials, colors and flavors.

Aseptic packaging

Aseptic packaging is an alternative to conventional canning in the production of shelf stable packaged food products. Aseptic packaging technology is fundamentally different from that of traditional food-processing systems. Canning processes commercially sterilize filled and sealed containers, while in aseptic packaging the presterilized product is filled into sterilized containers that are hermetically sealed in a commercially sterile environment. This form of heat treatment enables the commercial sterilization of the product, and results in minimal loss of product quality and nutrition. Various methods for the sterilization of packaging materials are currently used in aseptic packaging systems. Hydrogen peroxide sterilization followed by hot air appearsto have the most potential for use as in-line sterilization of packaging materials. Sterilization of packaging materials followed by ultraviolet irradiation has also been accepted for industrial application. Ultraviolet irradiation promotes the breakdown of peroxides into hydroxyl radicals, and the overall lethal effect is greater than the sum of the effects of peroxide and irradiation alone.

On-the-go Packaging

Owing to consumer's preference for easy and quick consumption of food while they are on the move, the packaging industry is focusing on the go packaging solution to meet the augmented demand for packaged and processed foods. The global on the go packaging market is expected to have a steady growth over the forecast period attributed to the increasing demand for easy usage and consumption of products. Growing affinity of the consumers for portability, easy usage, re-

On-the-go-packaging

Source: https://www.packagingtechtoday.com/markets/food/on-the-go-packaging/

usability, and re-closable characteristics is expected to boost the global on the go packaging market.

(a) Sustainable packaging (b) Portable Packaging (c) Transparent packaging

(*Source*: https://bizongo.com/blog/food-packaging-trends/)

Future trends in food packaging

A continuing trend in food packaging technology is the study and development of new materials that possess very high barrier properties. The use of high-barrier packaging materials reduces the costs in material handling, distribution/ transportation and waste reduction.Convenience is also a "hot" trend in food packaging development. Convenience at the manufacturing, distribution, transportation, sales, marketing, consumption andwaste disposal levels is very important and competitive. Convenience parameters may be related to productivity, processibility, ware housing, traceability, display qualities, tamper-resistance, easy opening and cooking preparation. A third important trend is safety, which is related to public health and to security against bioterrorism. It is particularly important because of the increase in the consumption of ready-to-eat products, minimally processed foods and pre-cut fruits and vegetables. Food-borne illnesses and malicious alteration of foods must be eliminatedfrom the food chain.Another significant issue in food packaging is that it should be natural and environmentally friendly. The substitution of artificial chemical ingredients in foods andin packaging materials with natural ingredients is always attractive to consumers.Many ingredients have been substituted with natural components – for example, chemical antioxidants such as BHA, BHT and TBHQ have been replaced with tocopherol and ascorbic acid mixtures for food products. To design environmentally friendly packaging systems that are more natural with high strength and good barrier properties is again a must.

References

Cerqueira, Miquel Angelo et al., 2015. Edible food packaging: materials and processing technologies. CRC Press https://1lib.in/book/2836247/887fef

https://b-ok.org/book/2278203/a48768

https://foodscience-technology.com/wp-content/uploads/2020/07/Innovations-in-Food-Packaging-Food-Science-and-Technology-International-1.pdf

https://www.appleacademicpress.com/innovative-packaging-of-fruits-and-vegetables-strategies-for-safety-and-quality-maintenance/9781771885973

https://www.futuremarketinsights.com/reports/on-the-go-packaging-market

Jung H. Han., 2005. Innovations in Food Packaging. Elsevier Academic Press, UK

Jung H. Han., 2014. Innovations in Food Packaging-2. Elsevier Academic Press, UK

Kerry J and Butler P., 2008. Smart Packaging Technologies for Fast Moving Consumer Goods. Wileyhttps://b-ok.org/book/567065/bd9303?dsource=recommend

Richard Coles, Mark J. Kirwan, 2011. Food and Beverage Packaging Technology. John Wiley & Sonshttps://b-ok.org/book/1165028/0f373e?dsource=recommend

Siddiqui, M. W., Rahman, M. S. and Wani, A. A. (Eds.). 2018. Innovative Packaging of Fruits and Vegetables Strategies for Safety and Quality Maintenance. Apple Academic Press, UK

Vigya Mishra, Neeru Dubey and Simple Kumar, 2018. Nano-Enabled Packaging of Food Products. Innovative packaging for fruits and vegetables-Strategies for safety and quality maintenance. Apple Academic Press.

16

Flavour in Sensory Science

M. Preema Devi[1], Dev Raj[2] and V.K. Joshi[3]

[1]*Department of Pomology and Post-Harvest Technology, Uttar Banga Krishi Viswavidyalaya, Pundibari, Cooch Behar, W.B.*

[2]*Department of Post Harvest Technology, ASPEE College of Horticulture, Navsari Agricultural University, Navsari, Gujarat-396 450, India*

[3]*Department of Food Science and Technology, Dr Y. S. Parmar University of Horticulture, Nauni, Solan, Himachal Pradesh*

Introduction

How one can differentiates between sensory science and sensory evaluation? Perhaps the simplest way would be to think of sensory science as the fundamental physiological and psychological underpinnings of sensory evaluation. It could include our basic understandings of the mechanisms of taste and odour, kinesthetic and the various techniques of scaling and statistical analysis, whereas, sensory evaluation could designate the application of this knowledge and the development of specific methods for the evaluation of the discriminative, descriptive and affective nature of products, in a problem-solving situation. When the quantitative descriptive analysis (QDA) methodology was first described, it represented a decade of experimentation and study of the issue that were perceived at the time to be impediments to solving "the problem of flavour characterization" (Stone *et al.*, 1974). As noted by Pangborn (1964), new developments and knowledge about sensory evaluation were beginning to have a positive effect on its acceptance by industry, but much more needed to be done if it was to contribute in a meaningful way. Today, it is indisputable that flavour is an important attribute affecting consumer acceptance of food. "Flavour is taste plus odour". The perception of taste is merely the detection of non-volative tastants by the tongue. On the other word, the perception of odour is the detection of odoriferous volatile aroma compounds by the olfactory nerves ending in the nose. The combination of all possible flavour stimuli we could encounter is as vast as the galaxy. However, most of the attempts to define these experiences have historically been over simplified.

Understanding Flavour Interactions

Development of a foundation for understanding relevant to flavour interactions must touch on the areas of:

- Percieved flavour
- The chemistry of flavour
- Some physical chemistry tools
- Flavour released during eating
- Food factors affecting flavour

Perceived Flavour

Flavour is a perceived attribute that results from our integrated response to a complex mixture of stimuli on several senses including smell, taste, touch, sight and even pain (Lawless and Lee, 1993). The process by which these responses are integrated is little understood partly because such a broad scope of disciplines, ranging from physico-chemical to biological to cognitive/ psychological is involved. Basically, aroma perception begins when volatile molecules are sniffed by the nose directly or flow to it retro nasally (via the backwards pathway from the mouth to the nose) during eating. When the molecules reach the olfactory bulb, a flavour signal is fired to the brain (Fig). The interpretation of these flavour signals depends on the taster and the context. Behavioural response to these flavours can vary within and between individuals depending upon attitude, memory, expectations, experience and physiology (Land, 1996; Acree, 1993).

The Chemistry of Flavour

The concept of chemicals as flavour stimuli is more than 170 years old (Brillat-Savarin, 1886). Though taste stimuli are important, the part of flavour that distinguishes different foods can be largely attributed to the volatile aroma molecules sensed by retro-nasal olfaction or the backward pathway that volatiles travel from the mouth to the nose during eating (Land, 1996).

Various techniques for determining which molecule are flavour active have emerged over the last decade (Grosch, 1993 and Acree *et al.*, 1984). Flavour-active molecules are plethoric in their sensory properties. Some are character impact compound i.e. those that when smelled, are clearly reminiscent of the named flavour (Reinecives, 1994). An example is benzaldehyde for cherry. For many foods, however, there is no known character impact compound. Instead, the proper balance of many flavour molecules is theorized to be key in providing the characteristic impression (Kosikowski and Mocquot, 1956).

Flavour molecules span a variety of chemical classes creating challenges for isolating and analyzing the molecules. Flavour molecules also exhibit different physical properties, behaviour and thresholds. Most are active at levels on the order of parts per million (ppm) or even parts per billion (ppb) and their thresholds can vary by as much as a factor of a billion. The tremendous dynamic range also contributes to the challenge of flavour analysis.

So, flavour are comprised of various aroma chemicals and like other chemicals, these will react or escape from the medium. Positive or negative changes can occur as a result of growth or maturation of the plant or animal source, cooking or other processing and interactions with the food matrix or package. During storage, reactions such as oxidation, hydrolysis and vapourization can occur, so that desired attributes may be lost and development of off-flavour can occur. Off-flavour compounds are flavour too and their behaviour is subject to the same influences as desired flavours.

Physical Chemistry Tools

Flavour equilibrium is never actually reached during eating but physical chemistry provides tools useful for understanding and predicting flavour behaviour. Flavour in bulk (vs. dispersed or emulsified) air, water and lipid phases and their combinations is reasonably well explained by equilibrium partitioning theory (Land, 1996). Specifically, the way flavour molecules distribute between air and water in a closed system at equilibrium is described by PC a/w, or the air/water partition coefficient.

A corresponding air/oil partition coefficient PC a/o (Catellan, 1971) described low flavour molecules distribute between air and oil. These coefficients indicate the amount of a flavour compound in the vapour phase (air) and as such are a useful measure of its availability to be sensed. While some air/liquid PC values for 25^{0}C are available in the literature, it may be necessary to measure values relevant to body temperatures, as these partition coefficients are quite temperature sensitive (Fisher *et al.* 1994). The higher the PC a/w in the oil/water coefficient that predicts how a flavour molecule partitions between oil and water phases, the more lipophillic the compound. A PC a/w of 1 indicates equal concentrations in oil and water. Most flavour compounds are at least moderately lipophilic (Table 1) and lipophillicity increases with molecular weight within a homologous series (Fister *et al.*, 1994). Compounds differ in concentration that vapourizes into the air based on their physical chemical properties. The medium too has a great effect on the concentration of flavour available to us in the vapour phase (de Roos, 1996). The data in Table 2 illustrate the effect of 3 different media on three compounds that is more volatile in oil than water. Vanillin on the other hand, experiences little media effect and might therefore, be considered nearly ideal in the sense of weathering matrix reformulations.

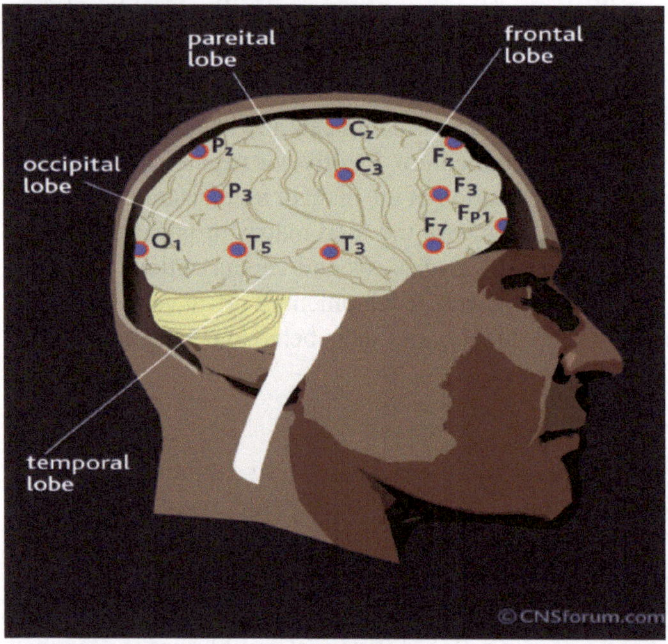

Flavour Release During Eating

To contribute to flavour perception, flavour molecules must first be released from the food. Both human and food factors affect this process. The human factors includes mastication habits, amount and composition of saliva, and other physiological characteristics that can vary from individual-to-individual.

As food is rarely in the mouth for longer than 20 to 30 sec., equilibrium with air or saliva is unlikely to be reached before swallowing (Land, 1996). Released of flavour during eating is a dynamic or kinetic process. The rate of flavour release in the mouth then, determines the extent of the total possible concentration that is reached (de Roos, 1996). Mc Nulty (1987) and Overbosch (1991) have extensively reviewed dynamic flavour release, concluding that the rate of release depends on both the compound and the characteristics of the medium. The type of molecules released and their quantities are key to perceived flavour.

Food Factor Affecting Flavour

The primary food factors are the composition and structure of the food. But the main factors affecting flavour in food are flavour-ingredient interactions, lipid flavour interactions, carbohydrate flavour interactions, protein-flavour interactions etc. knowledge on the profiling of flavour. There are two methods of flavour profiling.

Flavour Profiling Methods

Accepted approaches generally rely on only two parameters: impact and stimulus (Fig.1). At first glance, this seems to be a reasonable approach, and the resultant graphical depiction manifests itself as the typical spider-web configuration (Fig. 2). To expand the variables to be tested, the sensory analyst needs only to increase the number of spokes on the sensory wheel or points on the spider web. Although it is a commonly used and accepted methodology for flavour description, this approach ignores one crucial variable and can in some instances, be quite misleading, especially when the food product developer needs to assess what is really going on in the full scheme of the taste experiences. This variable is time.

When we taste a product, a multitude of aroma volatile race up the back of our throat towards the nasal cavity. These odouriferous gases fight for the opportunity to stimulate nerve ending in the soft tissue known as the olfactory bulb. The variety of aroma components in any given substances can number in the hundreds. When we look at a static spider-web profile we are ignoring strength vs. proportion, anesthetization, multi component synergism, masking, and most important, time. Lighter volatiles come-off first. Heavier volatiles are generally perceived later on which is totally ignored in the static model.

Some chemicals support each other, an effect called synergism while other chemical tend to mute others, an effect called as masking. Let us explore this issue in light of our simple system and how it might affect perception *vs.* time and therefore, skew a static analysis. Let us imagine a slightly more complicated system with a third component that comes out in the middle of the flavour/odour profile where it synergizes with the lighter ethyl acetate and masks the heavier vanillin. Such a compound is 6-methyl cocumarin, a creamy coconut-like material. As shown in Fig 3, the third component can change the profile and if the third component has its own character and dominate the middle ground profile, it becomes clear that midway in the profile a third unique flavour note would be apparent. Fig 4 illustrates this point appropriately.

Now picture scores, if not hundred, of volatile chemicals, each with its own characteristic vapour pressure and desensitization, synergism and marking characteristics, alone and in combinations with each other, and one can see that the resultant flavour profile can be extremely complex and dynamic. So, it can be concluded that in any multicomponent flavour system, the magnitude of the perception of a particular character or flavour nuance can often change within the brief tasting experience.

This is why an approach which utilizes a time – intensity curve is far more accurate. This approach addresses changes in characters by the addition of notations and 'ticks' in the curves where the profile alters appreciably in character. Fig 5 shows a typical time intensity curve.

The interesting magnitude y-axis seems to be a quantitative measurement, but there are too many simultaneous factors to make measurement precise. So, this approach addresses the subjective nature of the flavour profile and can be utilized to describe quality as well as provide a semi-quantitative objective measurement of strength. Because of this, the approach is best suited for comparative descriptions when measuring the change from a current product, deviation from a standard, or closeness to a target. It is also useful when a narrow wing of parameters or a windowing technique, is employed.

Dynamic Flavour Profile

The two types of flavour profile approaches look at only two dimensions of flavour at a time; the spider-web or descriptive attributes *vs.* intensity and the flavour profile curve addresses intensity *vs.* tasting time. Both are necessary tools. Since what is really going on in the taste experience is a continuing change of profile, the true measurement of flavour experience can best be accurately depicted by a graphic of continuing, *albeit* ever-changing, web profiles. This graphical depiction of a food product combining both approaches is called the Dynamic flavour Profile (DFP). As in other descriptive approaches, the DFP method requires trained panelists familiar with the nomenclature involved. A simple approach is used to flavour description terminologies which have its vocabulary based on similarities in chemical structure. It involves the theory that similar flavour types are generally similar in structure and can, therefore, be mixed as standards, each representing a different aroma type.

Applications of Dynamic Flavour Profile (DFP)

The following examples illustrate how the DFP approach can be applied to address various problems:

1. The product profile dissipates at the end upon the addition of another ingredient. The ingredient imparts a desired characteristic elsewhere, but not in the background. Therefore, its elimination is not recommended.

   The first choice is to modify the ingredient so that its effect on the product profile is minimized or eliminated. If the ingredient has a flavour itself, profiling it might help in determining what adjustments are feasible. Another approach might be to determine which of the 14 attribute have been affected by the addition of this ingredient. A before and after comparison of the DFP curves will illustrate this. By pinpointing the specific characteristics or nuances that are lacking, we can make adjustments by adoling these nuances separately or modifying a flavouring, if present, to augment those notes.

2. A product is modified to reduce the fat content. A change in profile is observed in the top note and a loss of overall creaminess is detected in the middle.

   One approach might be to determine the effect of different types of vegetable oil or oil replacers on the product. The DFP curve for each variable can be analyzed and the oil with the closest overall profile to the original could be selected.

3. When the product is processed, an off-note develops.

   The DFP method can be used as stated above, to select an additive that appears at the same time as the off-note and masks it.

4. The product flavour fades upon extended shelf-life, but previous direction to suppliers has been fruitless. The problem lies with the description of what is going on, and past attempts to do accurately have been frustrating.

   Since the 10 odours types used in the DFP method are analogues to chemical structure and terminologies commonly used by flavourists, pinpointing specific time-oriented and component-oriented descriptors should make the job as defined and targeted as possible.

5. Fruitiness of the product has been reduced. Example has shown that fruitiness is usually associated with esters. Since esters are volatile, it seems that the heat added to the product has to be minimized.

However, all attempts to address the problem in this manner have been of no avail. The DFP method determines that the ester profile is in fact high in front but lower in middle. This seems to be the area that needs support. Therefore, fruit flavours that contain middle volatile esters (amyl C-2, ethyl C-5 to C-7), such as apple, pineapple and banana flavours would be appropriate choice to minimize the problem.

Electronic Nose

Since, the graphical depiction of the DFP is somewhat complex, the later developments in computer technology come at the most opportune time. One such development is the electronic nose. One may think that an electronic nose would tell you everything you ever wanted to know about a sample but, in fact, an electronic nose functions like a black box-it knows nothing until it is taught. It is simply a series of sensors that respond to volatile components of the headspace above a sample (Michael *et al.*, 2000).

The experimenter optimizes operating conditions and teaches the electronic nose what to recognize through teaching sets and data libraries. The software of the electronic nose uses a series of algorithms and chemometric methods to provide meaningful data from the sensor response. Because this is a correlation technique, one must remember that the electronic nose will only be, at best, as accurate as the analytical or sensory data it is correlated against.

In the recent years, significant interest in the use of sensor arrays to discriminate between odours has arisen. If an array of non-specific sensors could be compiled to rival the human olfactory system, then the samples need not be separated and could be monitored analytically as a whole (Hardner and Barttert, 1992; Kress-Rogers, 1997).

The term "electronic nose" is applied to such an array of chemical sensors, where each sensor has only partial specificity to a wide range of odorant molecules, coupled with a suitable pattern recognition system. A sensor array used for discrimination between odors was first demonstrated by Persaud and Dodd in 1982. Subsequent improvement in both sensors and data analysis methods has led to the appearance of several commercial electronic noses.

Principles of Olfaction

To appreciate the operation of an electronic nose, it is necessary to have a basic understanding of the principles behind the human olfactory system. Fig. 8 illustrates the basic components of the human olfactory system and relates these to the corresponding components of an electronic nose. The sensation of smell is dependent upon the interaction of odorant molecules with a group of specialized nerve cells. The olfactory receptors, which are situated just above the bridge of the nose out of the main air stream. The hydrophobicity of a molecule is important, since the first step in the process of olfactory recognition is the dissolution of the molecule in an aqueous mucous layer covering the olfactory receptor cells. Each olfactory cell has a number of cilia, effectively increasing the surface area, which contain different guanine nucleotide binding proteins (Breer, 1994, Clapham, 1996). The genes encoding the proposed olfactory receptor proteins have recently been identified (Axel, 1995). The significances of the smell are illustrated by the sheer weight of genetic information used to control it. About 1,000 genes encode 1,000 different odor receptors. Thus, the odor receptor genes account for approximately 1% of all human genes. With regard to olfactory receptor structure, each protein is thought to transverse the cell membrane seven times, forming a pocket like structure into which the odor may bind specifically (Brur, 1994).

Many theories, such as the stereo-chemical theory of Amoore (1970) and the vibration theory of Wrigh (1982), have been proposed to explain the process of olfactory recognition. The exact method of interaction between an odor molecule and a receptor site has not been proved conclusively, but it is known that interaction results in excitation of the receptor cell. This produces a cascade of reactions which activity control in channels, within the cell membrane and results in an electrical signal that is passed along the axon to the olfactory bulb. Two transduction pathways have been identified using separate second messengers i.e., cAMP and IP3, which are thought to act in opposition to each other (Breer, 1994). Fig.10 illustrates the proposed mechanism for the generation of an electrical signal along an olfactory neuron.

The human olfactory system is quite remarkable in its ability. The number of distinct types of binding proteins is small (about 1,000), while the number of olfactory cells is large (about 100 million). The same proteins must be present in many different olfactory cells. Humans can detect at least 10,000 odors; therefore, each odor must bind to several receptors, producing receptors which are described as having partially overlapping sensitivities. However, the number of identifiable odours is far less, approximately 50.

The olfactory bulb is composed of three main layers: glomeruli. Thus, stimulation of a combination of olfactory receptors results in the formation of a two-dimensional topographical map representative of a particular odor. The electrical signals produced are further processed by the mitral cells and finally sent via the granular layer to the brain. The overall function of this stage is to reduce the noise associated with the signal and amplify it, effectively increasing both the sensitivity and the selectivity of the system. The electronic nose system parallels the human olfactory system in the following manner: Each chemical sensor represents a group of olfactory receptors and produces a time dependent electrical signal in response to an odor. Any noise and sensor drift may be reduced using signal preprocessing techniques. The final stage in the human olfactory process is the cerebral cortex of the brain, which classifies and memorizes odors (Schild, 1990); the equivalent process in the artificial nose is the use of pattern recognition software.

So, the advent of artificial sensory systems able to minimize chemical senses, such as those of electronic noses, has opened up a variety of practical applications and new possibilities in many fields where the presence of odors is the phenomenon under control (Kress Rogers, 1996). Many areas in the future will be touched by these noses, mainly owing to their promising, if still unknown, potential in terms of sensitivities, resolutions and stability and a huge growth of new market is foreseen. Typical examples are found in the field of food analysis, where sometimes very intrusive classes are adopted according to accepted categories such as freshness or edibility (Barttett *et al.*, 1997).

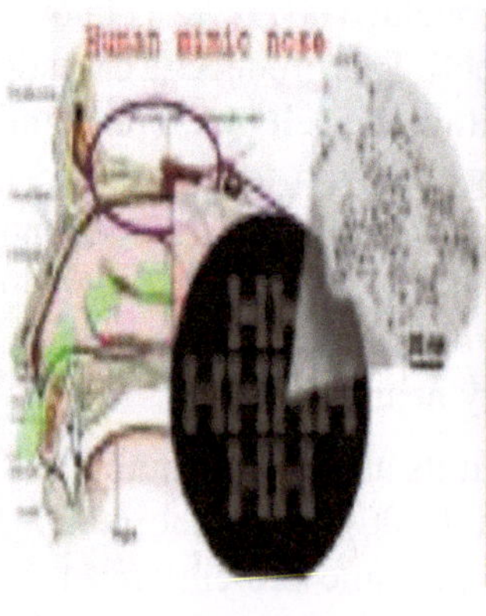

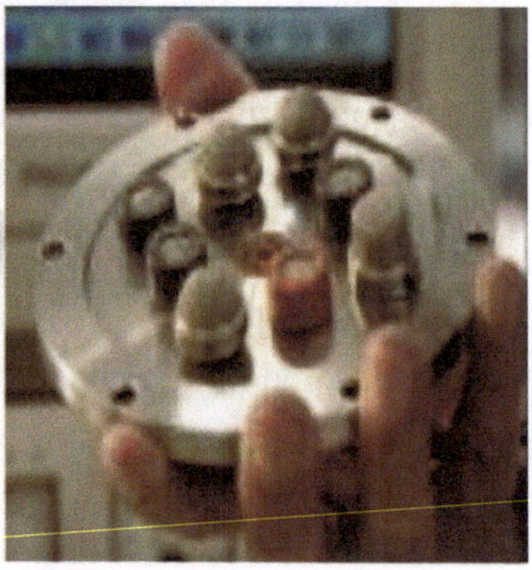

Application of Electronic Nose

1. Electronic nose for monitoring maturity and shelf-life of tomatoes:

   An electronic nose (E-nose) was used to evaluate the maturity and to monitor the shelf life of tomatoes. Results showed that pink stage fruits could be distinguished from light red stage and red stage ones based on the E-nose using principal component analysis (PCA) and linear discriminant analysis (LDA). The E-nose was able to classify samples having different firmness using PCA. By means of partial least square (PLS) based on E-nose sensor response signals, fruit firmness was predicted and the correlation between the predicted and measured values was 0.936. A clear distinction of initially light red fruit during storage from Day 1-6, Day 7-11 and Day 14-17 was obtainable by the electronic nose using PCA and LDA (Jun and Yibin, 2007). Research has shown that the E-nose has sufficient sensitivity and resolution to evaluate tomato fruit maturity and to product firmness. The shelf-life of light red stage and red stage fruit can be satisfactorily evaluated also.

2. Correlation between the electronic nose response and head-space volatiles of Tongkat Ali (*Eurycoma longifolia*) extracts:

   Most herbs have their own characteristic smell due to the presence of volatile compounds. Traditionally, volatiles are analyzed using sophisticated and expensive gas chromatography (GC) in tandem with selective mass spectrometric (MS) detector. An alternative approach has on the use of electronic nose is described. The approach is much simple than the traditional GC-MS counterparts, but providing key information of the samples analyzed. Using lipids and gas chromatography stationary phase materials with different polarity as sensing membrane, a quartz crystal microbalance smell sensor array has been developed for the analysis of traditional medicinal plant. The headspace vapour of different types of Tongkat Ali (*Eurycoma longifolia*) extracts were analyzed by the smell sensor and GC-MS. Correlation between the sensors response and identified compounds were studied using PCA. (Shafiqul Islam *et al.*, 2005).

3. Correlation of sensory analysis with electronic nose data:

   Gutierrez-Osuna *et al.*, (2001) had evaluated on the electronic nose as an alternative to sensory analysis for assessing the effectiveness of biofilters. An aroma San@ A32S electronic nose and a human panel at Duke University's Taste and Smell Research Lab were used to measure typical volatile compounds from swine confinement buildings. Chemo metrics techniques were employed to predict the olfactory scores of the human panel from the electronic nose data. The cross-sensitivity of the sensor array to the humidity of the samples is described. Results indicate that the E-nose generates responses that are correlated with sensory analysis rating of swine malodors, at different concentrations.

4. Forming odour categories using an electronic nose:

   The ability to use linguistic concepts to describe perceptions of measure data is an emerging feature for artificial sensing systems. The problem of symbolically representing odour categories using the data from an electronic nose. One objective is to facilitate human and computer interaction and therefore, the name given to the odours are correlated with a human user. Perceptual differences that arise between the human perception of odours and the electronic one, represents a challenge to the system. Therefore, to cope with these differences, the system maintains the freedom to evaluate how appropriately the linguistic concepts represent the sensory perceptions.

Finally, some experimental results are shown where the odour categories are formed and new odours are described using these categories.

5. Quartz resonator sensor arrays:

   Nakomota *et al* (1993) successfully identified perfumes and flavours by using quartz resonator with 8 different types of resonator membrane system initially designed to inspect which aroma. Both neutral networks and principal components analysis were used to aid data separation. Five different perfume types were clearly separated using PCA. Nato *et al.* (1995) showed that an epoxy coated quartz crystal resonator sensor, used with associative neutral network analysis and principle component analysis, could discriminate between red, white, and rose wine. Nine parameters representative of the transient sensory response curve were chosen and analyzed. The neural network after initial training with 5,000 known samples, was able to identify each kind of wine with a 100% success rate.

Conclusion

The quality of the food supply depends at least in part on the sensory effects that foods have on consumers. Food acceptance is not just a marketing or product development issue; it has broad health and nutritional implications. Consumers are influenced by the sensory effects of food in market purchase decision regarding all food products, including those of health benefits or risks. Sensory research has played a key role in the development of our modern food supply and will continue to do so as long as people consume food and beverage.

Modern developments in terms of flavour analysis in sensory science have been observed liked the Dynamic flavour profiling and the electronic nose. Electronic nose application is widespread but many problems have been encountered while trying to develop an electronic nose suitable for all applications. Problems associated with sampling, calibration, sensor drift and control of humidity and temperature are the major causes for concern. Improvements in associated technologies such as new odour - sensing materials, development of hybrid noses and smarter pattern recognition techniques are essential if the market potential of such an instrument is to be realized. Ultimately, the development application-specific noses that are tailor made for a given problem may prove more successful than attempts to develop a single system suitable for all applications.

References

Acree T.E., Barnard J. and Cunningham D.G. 1984. A procedure for the sensory analysis of gas chromatographic effluents. Food Chem. 14:273-286.

Acree T.E. 1993. Bioassays for Flavour. In: Flavour Science – sensible principles and techniques, eds. T.E. Acre and R. Teranishi. pp. 1-20. ACS, Washington, D.C.

Aishima T. 1991c. Discrimination of liqueur aromas by pattern recognition analysis of responses from a gas sensory array. Aral Chin Acta. 243:293-300.

Aishima T. and Nahal S. 1991. Chemo metric of flavour research. Food Rev. Intl. 7(1): 33-101.

Amoore J.E. 1970. Molecular Basis of odour classification. C.C. Thosam, Spingfield, III.

Axel R. 1995. The molecular logic of smell. Sci. Am. Oct. pp. 130-137.

Barttett, P.N., Elliott, J.M. and Gardner J.W. 1997. Electronic noses and their application in the food industry. Food Technol. 51:44-48.

Breer H. 1994. Signal recognition and chemoelectrical transduction in olfaction. Biosensors Bioelectronics. 9:625-632.

Catellan G.W. 1971. Physical chemistry, 2nd ed., Addison – Wisley Publ., Reading, Mass.

Clapham D.E. 1996. The G-protein nanomachine. Nature. 379: 297:299.

De Roos K. 1996. Personal communications.

Dolf D.1996. The Dynamic flavour profile method. Food Technol. 2:55-60.

Fiorella Sinesio, Corrado D.N., Giovanni B.Q., Francesco M.B., Elisabetta M/., Antonella M., Reberto P. and Amaldo D.A. 2000. Use of electronic nose and drained sensory panel in the evaluation of tomato quality. J. Sci. Food Agric. 80:63-71.

Fischer N. and Widder S. 1997. How proteins influence food flavour. Food Technol. 51(1): 68-70.

Fisher R.G., Toui I.C., Bodett, S.P., Leland J.V., Croasmun W.R. and White J.D/. 1994. Effects of solids on equilibrium flavour releases. Presented at Ann. Mtg., Am. Chem. Soc., Washington D.C., Aug. 21-26.

Fisher R.G. 1994. Personal communication. Kraft Foods, Glenview, 12.

Gardner J.W. and Barttett. P.N. 1992. Sensors and sensory systems for an electronic nose. NATO ASI series E: Applied Sciences, Vo. 212, 161-179. Kluwer, Dordrecht the Netherlands.

Gardner J.W. and Hines E.L. 1997. Pattern analysis techniques in Kress-Rogers. Pp. 633-652.

Grasch W. 1993. Detection of patent odorants in food by aroma extract dilution analysis. Trends in Food Science and Technology. 4:68-73.

Groosch W.1993. Detection of patent odorants in foods by aroma extracts dilution analysis. Trends in Food Science and Technology. 4:68-73.

Gutierrez-Osuna R., Schiffman S.S. and Nagle H.T. 2001. Correlation of sensory analysis with electronic nose data for swine odor remediation assessment. Proceeding of the 3rd European Congress on Odours, Metrology and Electronic noses, Paris, France, June 19-21.

Jane V. Leland. 1997. Flavour interaction: the greater whole. Food Technol. 51(1): 75-80.

Jun W. and Yibin Z. 2007. Electronic nose technique: potential for monitoring maturity and shelf life of tomatoes. New Jealand J. of Agricultural Resarch. 50: 1219-1228.

Kosikowski F.V. and Mocquot G. 1958. FAO Agric. Stud. 38: 133-147.

Kress – Rogers E. 1996. Handbook of Biosensors and electronic Nose, CRC Press, Boca Raton, FL.

Land D.G. 1996. Perspectives on the effects of interactions on flavour perception: an overview In: Flavour – Food Interactions ed. R.J. Mc Gorrin and J.V. Leland pp. 2-11. ACS Symposium Series Book 633.

Lawless H.T. and Lee C.B. 1993. Common chemical senses in food flavour. In: Flavour Science-sensible principles and techniques, ed. T.E. Acree and R. Teranishi pp. 23-66, ACS. Washington D.C.

Maur H.G. 1975. Binding of volatile aroma substances to nutrients and foodstuffs. Proc. Itern. Symp. Aroma Res. Zeist. Pudoc, Wageningin. 143-157.

Mc. Nullty P.B. 1987. Flavour release – clusive and dynamic. In: Food structure and behaviour. ed.: Blaushard J.M.V. and Lellford O. pp. 245-258. Academic London.

Michael G.M. and Roman D. Grypa. 2000. Spices, flavour system, the electronic nose. Food Technol. 54(3): 44-46.

Nakomoto T., Fikada A. and Morllzumi T. 1993. Perfume and flavour identifications by odor-sensing system using quartz resonator array and neural network pattern recognition. Sensors Actiuators. B, 10: 85-90.

Nanto H., Tsubakino S., Ikeda M. and Fumitaka E. 1995. Identification of aromas from wine using quartz – resonator gas sensors in conjuction with neural – network analysis. Sensor Actuators. B, 24s-25: 794-796.

Overbosch P., Afterof W.G.M. and Haring P.G.M. 1991. Flavour Release in the mouth. Food Revs. Int. 7(2):137-184.

Pangborn R.M. 1964. Sensory evaluation of foods: A book backward and forward. Food Technol. 18:63-65.

Persaud K. and Dodd G.H. 1982. Analysis of the mammalian olfactory system using a model nose. Nature. 299:352-355.

Persaud K.C. and Pelosi P. 1995. An approach to an electronic nose. Artificial Internal Organs. 31:297-300.

Persaud K.C. and Travers P.J. 1997. Arrays of broad specificity firms for sensing volatile chemicals in kress-Rogers.

Philip N., Bartlett J.M., Elliott and Julian W. 1997. Electronic Noses and their application in food industry. Food Technol. 51(12):44-48.

Runeccuis G.A. 1994. The flavorist. In: Source Book of Flavour, Second edition. Ed. G.A. Runeccius, pp. 691-712. Chapman & Hall. New York.

Schild D. 1990. Chemosensory information Processing NAT%D ASI Series H: Cell Biology, 39. Springer-Verlag Berlin.

Shafiqul Islam A.K.M., Othman A.R., Ahmad M.N. and Md. Shakaff. A. Y. 2005. Correlation study between the electronic nose response and headspace volatiles of Eurycoma longifolia extracts. Ist International Conference on Sensing Technology. Nov. 21-23, Palmerston North, New Zealand.

Shafiqul Islam A.K.M., Othman A.R., Ahmad M.N. and Md. Shakaff. A.Y. 2005. Correlation study between the electronic nose response and headspace volatiles of Eurycoma longifolia extracts. I[st] International Conference on Sensing Technology. Nov. 21-23, Palmerston North, New Zealand.

Stone H., Sidel J., Oliver S.M., Woolsey A. and Singleton R.C. 1974. Sensory evaluation by quantitative descriptive analysis. Food Technol. 28(1): 24-34.

Wright R.H. 1982. The sense of smell. CRC Press. Boca Raton. FL.

17

Food Safety and Quality Assurance

Satish Kumar

Department of Food Science and Technology, Dr. Y. S. Parmar University of Horticulture and Forestry, Nauni, Solan Himachal Pradesh-173230, India

Introduction

Increased awareness about the food safety and various adulterations or crises reported in food industries, the food quality and its safety has emerged as the most important features of the modern food processing and preservation. The important historical dynamics which raised the concerns of food safety were motivated by accidents related to food borne pathogens and chemical contamination during the food processing and distributions. Many times, the terms food quality and safety are interchanged however, there is a difference in terms of their perception by the consumers and manufactures. In past few years, the food safety has become the characteristic of food quality and has been regulated by various national and international agencies. The increased concerns about the safe, nutritious and high quality food has over the time forced the public authorities to engage the food industries to construct the quality management systems in order to improve the food safety and build the consumers believe and trust towards processed foods. Food quality is a broader concept which includes food health quality attributes along with parameters like sensory characteristics like taste, smell and convenience like food preparation (Röhr *et al.*, 2005). The food qualities like nutritional, convenience and sensory are not requested by the public laws but are desired by the consumers. The management systems for the quality assurance have evolved over the time and now we have systems like HACCP. ISO 9000 and ISO 14000 in place which are well known quality management modules to ensure the elimination of food hazards. On a broader scale of trading food several technical agreements under WTO (world trade organization) have been made which provide equal opportunities to all countries to trade the goods however, it imposes the similar food quality standards across all the participatory members. The responsibility of food safety is directly connected with all the stakeholders of the food chain which include manufactures, processors, farmers, food handlers and finally the consumers. Food contaminations may be a result of poor manufacturing practices and poor storage and handling practices at any stage in the food chain. To ensure the safety compliances, national food regulatory models plays a significant role. The guidelines issued by FAO/WHO, 2003 are focused on the public organizations, as well as food control authority self-assessment to ensure highest food quality and its safety from farm to fork. The base of the assessment includes 6 factors viz. food inspection, food quality management, food safety, official food control laboratories, food legislation food and quality information, education and communication (Shukla *et al.*, 2014).For effective implementation of the food safety and quality module the countries have developed various food laws and legal norms which states the principles to be followed for raw materials, their processing and distribution in order to achieve the consumer expectations as well as the hidden food safety concerns.

Food Quality

Quality is the sum total of the attributes of a products which makes it what it is. It defines the superiority of a food product and helps the people differentiate between two units of same commodity. Food quality and safety are amongst the most discussed topics in past few decades by the food policy makers. There is a similar agreement between the subjective and objective dimension of quality. The subjective quality means the consumer perception about the product like its taste, color, texture which may vary from person to person. Whereas, objective quality includes different physical characteristics in order to manufacture the product which is deal of food engineers and technologists. Food quality is a holistic approach any may involve various factors as depicted in Figure 1

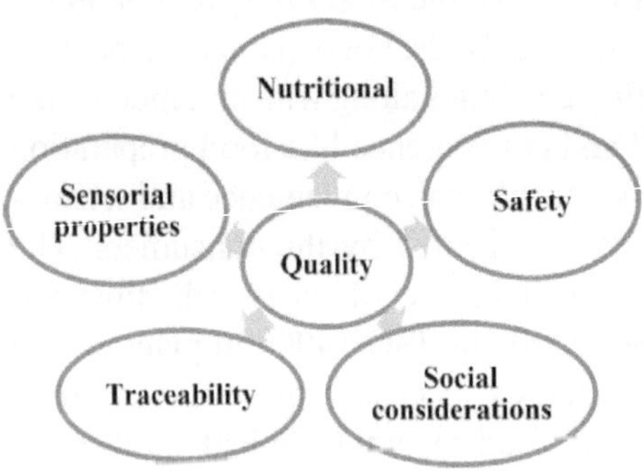

Fig. 1. Holistic, multi-factor concept of quality

The dimensions of food quality include performance, reliability, durability, serviceability, aesthetics, features, perceived quality and conformance to standards. There are several type of food quality approaches like the transcendent approach in which quality is viewed as innate excellence, the manufacturing-based approach which assumes quality is all about making or providing error-free products or services, the user-based approach assumes quality is all about providing products or services that are fit for their purpose and need, the product-based approach views quality as a precise and measurable set of characteristics and the value-based approach defines quality in terms of market valueThe food quality perception and its effect on consumers choice have be studied by using different approaches such as economics of information, end approaches, satisfaction or dissatisfaction approaches and expectancy value approaches. In consumer's point of view, quality of a food products are taste (sensory qualities), convenience, health and some process qualities like organic products, GMO-free, natural products. All these quality parameters may be affected by various factors to different extents and the regulations are gramed in such a way to minimise the interaction of the food with such factors which can negatively deteriorate the quality of the food. The main factors which can affect the food quality are given hereunder in Figure 2.

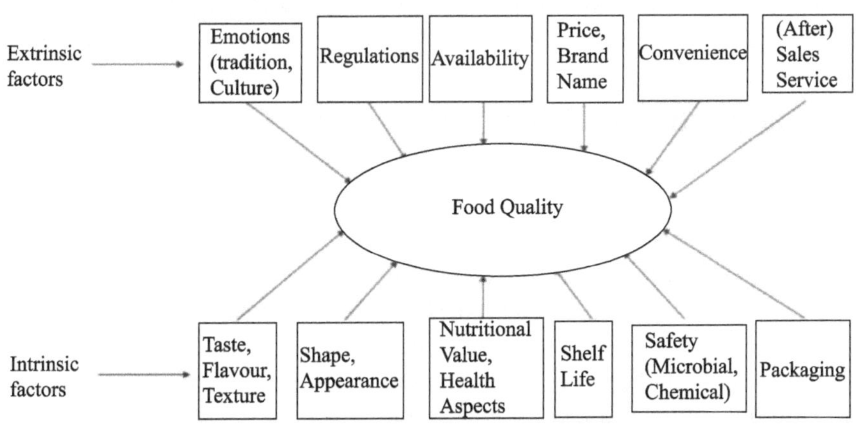

Fig. 2. Factors affect food quality

Quality Control v/s Quality Assurance

Rigorous efforts for the control of quality started early in the 20th century. The concept of quality control was a broader approach earlier in order to attain the quality from inspection to today's terms hazard prevention. Moving a step further the word control now also includes the quality planning concepts. But recently the word quality control has narrowed and is defined as the approach used to verify that deliverables are of acceptable quality and that they are complete and correct. The basic activity of quality control includes establishing the specifications of the parameters which are to be controlled;preparing the plans for quality control; checking the performing or inspections;diagnosing and taking action on the variations observed; andchecking that the variations have been corrected.The other term quality assurance is a function of total quality management. Quality assurance is a planned process used before the starting the production process and other related activities. The principle of quality assurance is useful when appropriate technical knowledge, scientific experience and training is established by producers. Quality assurance is modern term which describes the control, evaluation and auditing of various food processing system.The quality assurance gives the confidence to the management, consumers and customer.There is a very thin line of difference between the quality control and quality assurance inspite of being very similar to each other like both the terms focus on performance evaluation, compare performance goals. However, both the terms are different as quality control implies to maintain the control, performance during the operation and goal comparison whereas, quality assurance focuses on verifying the control which is being operated and maintained (Juran & Godfrey, 1999). Quality assurance is managerial tool whereas, quality control is corrective tool. Also, quality control is product oriented process whereas, quality assurance is the process oriented.

Total Quality Management (TQM)

Total quality management is defined as a process used to manage the overall quality of the food products. The TQM principles exists in the center of any business which ensures the utmost production and minimum cost. The aim of TQM is the highest customer satisfaction in the products delivered. It is an integration of all the activities and functions which exists within an organization to achieve quality goals continuously. The main pillars of TQM are the process and quality management, quality practice, customer focus and employee improvement. While, the elements of TQM includes meeting customer requirements, to do error free work, managing by prevention, measuring the cost of non-quality, never ending improvement (continuous improvement) and education and training (Psomas and Fotopoulos, 2010). Various components of TQM has been discussed in Figure 3.

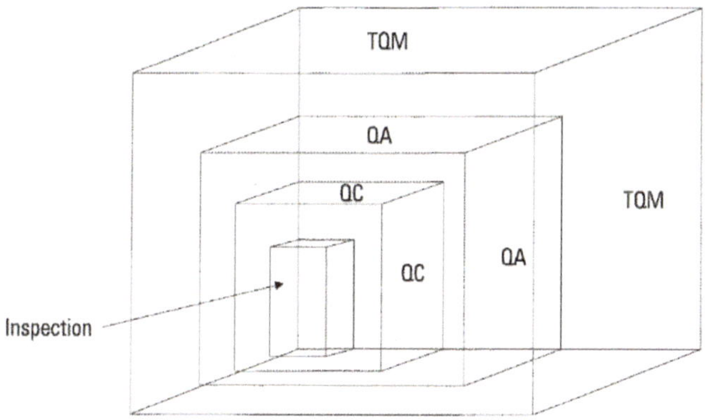

Fig. 3. Relationship between QA, QC and TQM

Hazard Analysis and Critical Control Point (HACCP)

Food safety has now emerged as a global concern which makes health related agencies and governments of many countries to seek for more effective methods to monitor the production processes. Hazard Analysis and Critical Control Point (HACCP) is a system which ensures the limits the risks of hazards occurring during food processing. Now it is internationally accepted system but initially it was introduced in United State in year 1971 by collaboration with Pillsbury Company and NASA (NationalAeronautics and Space Administration) and US Army Natick Research and Development Laboratories. In year 1972, the US based company started the application of HACCP for the manufacturing of food products. Later in 1990, globalization of HACCP started with meat and fish industries and was later approved by Indian government also. Food safety management system is defined as a set of interconnected elements which establishes the policies and objectives in order to achieve them by using controlled and direct organization to ensure the food safety correlated with concept of HACCP.

The development and successful implementation of HACCP plans starts with many prerequisite programs like GMP (good manufacturing practices), cGMP (current good manufacturing practices), GAP (good agricultural practices), GHP (good hygiene practices), GLP (good laboratory practices) and GDP (good documentation practices). HACCP principles are based on food safety system and have been applied successfully in different food processing plants, food services and retail food stores. There are total 7 principles of HACCP which is adopted globally by government agencies, food industries and trade association. HACCP principle is applied after the establishment of prerequisite programs like GMP and GHP to maintain the hygienic environment in manufacturing plants. The prerequisite programs are used in order to control hazards in surrounding and thereby prevent contamination of the processed products. Prerequisite program provide strength to the HACCP plans. The seven principles of HACCP are:

Principle 1: Conduct a hazard analysis.

Principle 2: Determine the critical control points (CCPs).

Principle 3: Establish critical limits.

Principle 4: Establish monitoring procedures.

Principle 5: Establish corrective actions.

Principle 6: Establish verification procedures.

Principle 7: Establish record-keeping and documentation procedures

Further there are 12 task are enlisted in HACCP for its effective implementation in the food industry. The 12 tasks are enlisted hereunder:

1. Establish a HACCP team
2. Describe the product
3. Identify the product's intended use
4. Draw up the commodity flow diagram
5. Onsite confirmation of flow diagram
6. Identify & analyze hazard(s) – Principle 1
7. Determine the CCPs – Principle 2
8. Establish critical limits for each CCP – Principle 3
9. Establish monitoring procedure – Principle 4
10. Establish corrective action – Principle 5
11. Verify the HACCP plan – Principle 6
12. Keep record – Principle 7

Good Manufacturing Practices (GMP)

Good manufacturing practices (GMP) are practices that follow the guidelines given by the organizations that control authorization and licensing for manufacture and sale of food and its products, drug products, and active pharmaceutical products of highest quality. GMP is a system that ensures that the food meets criteria like safety, quality and legal requirements. Application of GMP is involved in personal hygiene, air quality, raw materials, operation controls, lighting, water used in various stages of production and cleaning purposes, packaging, transportation, marketing, and so on. Ten principles of GMP are:

1. Writing procedure
2. Following written procedures
3. Documenting for traceability
4. Designing facilities and equipments
5. Maintaining facilities and equipments
6. Validating work
7. Job competence
8. Cleanliness
9. Component control
10. Auditing for compliance

GMP protects consumers from adulterated, mislabelling of product. Secondly, it helps to reduce cross contamination and poor sanitization and misidentification of ingredients; enhancing consistent industry-wide requirements and assurance to consumers on product quality. Moreover, at the time of emergency product recall can be easily achieved with the help of GMP. Similarly, implementation of GMP is necessary in a Quality Control Laboratories as it helps to prevent the errors and build a quality procedure or methodology.

Good Agricultural Practices (GAP)

According to United Nations Food and Agriculture Organization (FAO), GAP is defined as "practices that address environmental, economic and social sustainability for on-farm processes, and result in safe and quality food and non-food agricultural products". The four main pillars of GAP are:

- Economic viability
- Environmental sustainability
- Social acceptability
- Food safety and quality

Application of GAP in systematic methodology from farm to fork ensures improvement in safety and quality of food and agricultural products. GAP helps to protect food in primary stage of production from hazards, toxic materials, biological contaminants, etc. Additionally, producers and consumers will benefit from global markets and improve their livelihoods and the national economy as a whole. That is, from moisture, temperature, preparation of soil to replantation, harvesting techniques play vital role in promoting sustainable agriculture worldwide (Dudeja & Singh, 2018). The implementation of GAP has certain challenges as follows:

- Need of awareness: Both producers and consumers should have the correct knowledge foe the implementation of GAP so that better quality, yield and productivity can be obtained.
- Improper system to achieve goods according to the specified quality and safety standards.
- Sustainable environment, health and safety concerns for the workers.
- Awareness among small-scale farmers regarding the export market opportunities until and unless they are provided with adequate information.
- A proper framework is required for the implementation of food safety programme.

Good Hygiene Practices (GHP)

The acronym of GHP is good hygiene practices which is used in food industries and is essential for protecting the consumer and prevent several public health risks like foodborne illness, toxicities, and microbial outbreaks which may be due to the poor hygiene awareness among food workers. Nearly, 97% of total foodborne illness outbreak is due to the poor personal hygiene of food workers. Hence, it becomes necessary to discuss and create awareness among the entire workers who are occupied in food handling sectors. The awareness related to cleanliness of cloths and body, wearing protective and suitable cloths to ensure food safety and thereby public health. The unhygienic processing environment of food led to the contact with the harmful pathogens leading to the serious illness in consumers followed by outbreaks (Wambui *et al.*, 2017).

Good Laboratory Practices

The good laboratory practices include non-clinical laboratories that focus on the research or permits the products marketing to ensure the quality and integrity of studies being carried out. It is regulated by government authorities. By definition, GLP is a set of principles whichdevelops a frame work in which laboratory based studies are planned monitored,performed, and archived and lastly reported. GLP is mostly associated with many industries like pharmaceutical and food processing for development of new products. In food industries, GLP applied in food additives, food packaging, colour additives, food contamination limits, etc. GLP doesn't applicable for clinical studies, it only limited up to non-clinical studies only. For clinical studies, there is a separate system called good clinical practices (GCP).In the 1970's, Food and Drug Administration came to know about the poor laboratory practices and discovered lot of fraud cases in USA.GLPis then onwards regulated by FDA since 1978.

Good Documentation Practices (GDP)

The critical part of GMP programs in an organization lies within the effective management and control of documents. Maintenance of documents can never be an option as it's the legal need. Good documentation is a key to GMP as it keep eye on all possible testing activities, manufacturing and developments and ensures their traceability. Good documentation act as a route to conduct audits in order to analyse the quality of various unit operations within the manufacturing plant. Mostly, documentation includes records and processes. The records are generally attributed to standard operating procedures (SOP) and possess confirmed details which is requirement of SOP (Sandle, 2014).

Good Distribution Practice (GDP)

The good distribution practices includes the record keeping, information about products like storage, temperature, tracking ID, etc. along with cleanliness and organization in shipping process. Good receiving or shipping process includes one important policy which is FIFO i.e., first in first out. The distributors should ensure that the products arrives first should be shipped first and should not be stored for very long period of time. This will prevent the shelf life losses and wastage of expiring products and hence provide freshest products to the consumers.

Regulatory Framework to Ensure Food Quality and Safety

The methodologies involved for processing, preservation and packaging of food products are covered by numerous rules and regulations in various countries.Agencies and institutions have also been established to lay down standards for the quality of foods before reaching the consumers. Failure of a consumable product to meet the regulatory requirements such as standard of identity, the declared quantity and ingredients, or label claims, etc. can be considered as mis-representation, misbranding, or fraud. The 4 common standards are:

1. Legal standards
2. Company or voluntary label standards
3. Industry standards
4. Consumer or grade standards.

Effective means of food quality can be achieved by legislative measures, certification schemes and public participation and involvement in the programme.

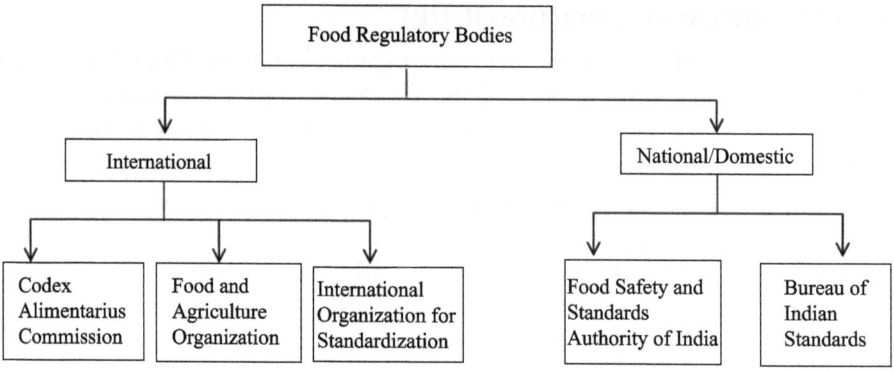

Fig. 4: International and nation food regulatory bodies

1. Food Safety and Standards Authority of India (FSSAI)

FSSAI has emerged as the most important and versatile system of food quality management and safety assurance in India. FSSAI was established in 2006 under the act of Food Safety and Standards Act, 2006 which defined food safety as an "assurance that food is acceptable for human consumption according to its intended use". Prior to this act, numerous laws and legislations existed under various ministries or organizations or agencies, which caused the conflicts among methodologies, procedures, quality, and standards of food product. Thus introduction of single law with apex regulatory body that deal with manufacturing, distribution, storage, sale and import of food products. Rules and regulations of FSSAI are extended all over the nation providing uniformity for licensing or registration regime. Some of the historic prominent acts and laws which are covered under FSSAI to maintain the quality of food includes:

1. The Prevention of Food Adulteration Act, 1954 (PFA)
2. Fruits and Vegetable Products (Control) Order- FPO 1955
3. Meat Food Products Order (MFPO)1973
4. Vegetable Oil Products (Control) Order, 1947
5. Edible Oils Packaging (Regulation) Order, 1988
6. Solvent Extracted Oil, De-oiled Meal and Edible Flour (Control) Order, 1967
7. Milk and Milk Products Order, 1992

2. Codex Alimentarius

With the increasing demand of food and food products, it has become necessity to ensure the health of consumers and fair trade practices. Further, with the rapid urbanisation and opening of the global food markets, the food from around the globe has become accessible. As a result, Codex Alimentarius Commission was developed in June 1962 by joining Food and Agriculture Organization (FAO) and World Health Organization (WHO) and its first session was held in Rome, 1963. Codex Alimentarius includes international standards, codes of practices, methodologies, etc. and production and safety related to food and its associated products. At present, Codex Alimentarius Commission (CAC) has more than 200 observers including 52 inter-governmental, 162 non-governmental and 16 United Nations organizations. Till date, establishment of codex helped to define and standardize the requirements of food products, which in turn facilitated international trade to take place harmoniously.

Codex Alimentarius includes various categories of foods viz. processed, unprocessed or raw which are intended for the consumption along with the materials involved in both preparation and treatment of food products. However, tobacco and other related substances which serve the purpose of drugs are excluded by CAC. Codex also regulates quality and hygiene related to food additives, microbiological standards, pesticides, food labelling, marketing, other hazardous substances like contaminants, residues of antibiotics and various veterinary drugs, etc. Regulations over safety methods while, involving most modern biotechnological methods in production, import-export of food, certification system, sampling techniques, hazard analysis, and so on are notable area of codex. Hence, regulation over all food and its related area brings out the safety measures, standards and uniformity of products and trade worldwide (Das, S*et al* 2020). According to Codex Alimentarius, detailed information should be labelled on the food products, so that it guides the consumers to select correct product. The information includes:

1. Name of the food/ product
2. List of ingredients have to mention indescending order
3. Net content and drained weight of the product
4. Name and address of manufacturer/company
5. Country of origin
6. Lot identification/number
7. Manufacture date and storage instructions
8. Instructions for use of the product

3. Food and Agriculture Organization (FAO)

Food security being the prime concern of many organizations across the world, FAO aims to create a zero-hunger-world ensuring regular access to nutritious food and thus a better and healthy life. FAO being a specialized agency of United Nations was established in the year 1945, have responsibility to coordinate government and other agencies in the programs aiming for the development of agriculture, fisheries, forestry, land and water resources. They gather information, followed by analysis, interpretation and finally disseminate inorder to achieve agriculture development and attain food security.

4. International Organization for Standardization (ISO)

ISO being the worldwide federation of many national standards bodies and these works are carried out by ISO technical committees. International organizations, governmental and non-governmental in alliance with ISO are involved in these activities. There are different set of ISO standards have been established for specific functions. Some of them are ISO 9000 aiming at quality management and ISO 22000 deals with food safety management. ISO does not provide certificates, neither carry out certification, nor control certification bodies; instead they develop standards and guidelines to encourage good practice for achieving accreditation and certification. ISO ensures the reliability, confidence in consumers by opening a legal and standardized world trade and expert opinions to government, extra money and time involved in the registration process, failure to provide product quality, etc. are some of its drawbacks (International Organization for Standardization, 2018).

5. Bureau of Indian Standards (BIS)

Standards are laid for numerous products and services including vegetable and fruit products, spices and condiments, animal products, processed foods, etc. BIS was established under the Bureau of Indian Standards Act, 2016 and identified as 'National Standards Body of India'.Even though BIS marking is mandatory of products, commodities that have influence on the health and safety of consumers such as baby foods, electrical appliances, etc. should contain ISI (Indian Standards Institute). The products are checked for quality by the ISI in their own network of testing laboratories at Delhi, Bombay, Calcutta, Madras, Chandigarh and Patna or in a number of public and private laboratories recognized by them.BIS helps the nation's consumers to evaluate and identify the quality products as the agency itself has given prime importance in standardization and conformity assessment.

6. Agricultural Produce Grading and Marketing (AGMARK)

A word AGMARK is a derived from Agricultural Marketing and was set up by the Directorate of Marketing and Inspection of the Government of India by introducing an Agricultural produce Act in 1937. The word 'AGMARK' seal ensures quality and purity. The commodities are graded according to the quality, which helps the farmers or traders to identify the products easily. Consumers can also distinguish the products as they desire. Numerous commodities such as pulses, oilseeds, cereals, honey, fruits, vegetables, vegetable oils,butter, ghee, spices, etc. are included under AGMARK act. Some of the advantages are ensuring the stability in price of products in market, helps to facilitate inter-state and international marketing, helps to improve agriculture and its technologies, improves the trust within the consumers, and so on.

7. Intellectual Property Rights (IPR) related issues

Intellectual property rights (IPRs) are legal rights governing the use of creations of the human mind and have been recognized as an asset and means of rewarding and harvesting the fruits of research and development. Geographical Indications, Trademark, Trade secret, protection of plant varieties, patent, copyright, design registration, etc. have high relevance in areas like food, agriculture, and their allied sectors. World Intellectual Property Organization (WIPO) established by the WIPO convention in 1967, is a specialized agency of the United Nations, and have its head quarters at Geneva. This organization helps in economic development, stimulates innovation, safe guards the public interests and rewards the creativity. On the other hand, World Trade Organization (WTO) – TRIPS Agreement deals with the protection & enforcement of "Trade-Related" intellectual property "rights". It is an attempt to narrow the gaps in the way these rights are protected around the world, and to bring them under common international rules (Savale & Savale, 2016). A trademark or "brand name" comprises "anything" that identifies the source of goods or services from those of another when used in commerce. It helps (i) to identifies the goods /or services andits origin, (ii) to guarantees the stable quality and (iii) to advertises the goods/services. Apart from this, Geographical Indication is given importance as they provide an indication which identifies goods as originating or manufactured in the territory of country or a region or locality in that territory where a given quality, reputation or other characteristic of such goods is essentially attributable to its geographical origin, for instance, Mizo chilli from Mizoram and Bangalore Rose Onion (*Gulabi Eerulli*) from Karnataka.

Conclusion

Food and food products have direct involvement in the health aspects and overall well-being of the consumers, regardless of their age and genderand hence, should be safer and of the highest nutritional and functional quality. Due to perishability, chances of cross-contaminations, etc. probability of spoilage and presence of other hazardous materials at all stages it becomes utmost important to continuously monitor the production, packaging, storage and transportation of the foods for their quality and safety. The introduction ofvarious laws and regulations to monitor different unit operations from the food production till its distribution worldwide has resulted in consumer confidence about the higher quality and safety of the food. Additionally, systematic procedures and methodologies under the guidelines of national and international regulatory bodies helps produces, manufacturers, sellers and buyers at various levels-from identifying the product grade/quality to choosing reliable products, and so on. Moreover, the regulations related to intellectual property rights, especially patents, geographical indications, etc. are vital for protecting the new inventions and/or creations of people while ensuring the profits to the stakeholders. IPR linked pathway may facilitate the growth of agri- business and industries in the near future while, the consumer driven processing industry is also transforming the food products to retain the market with continuously changing food habits and highest quality and safety requirements of the consumers.

References

Das, S., Das, S. S., & Chowdhury, S., 2020. Codex Alimentarius and its Role. Biotica Research Today, 2(4): 54-56.

Dudeja, P., & Singh, A., 2018. Good Agricultural Practices. In Food safety: Farm to fork implementation. CBS Publishers New Delhi, 14: 118-120.

International Organization for Standardization, 2018. Food Safety Management Systems: Requirements for Any Organization in the Food Chain. ISO.

Juran, J., & Godfrey, A. B., 1999. Quality handbook. Republished McGraw-Hill, 173(8).

Psomas, E. L., & Fotopoulos, C. V., 2010. Total quality management practices and results in food companies. International Journal of Productivity and Performance Management.

Röhr, A., Lüddecke, K., Drusch, S., Müller, M. J., & Alvensleben, R. V., 2005. Food quality and safety-consumer perception and public health concern. Food control, 16(8): 649-655.

Sandle, T., 2014. Good documentation practices. Journal of validation technology.

Savale, S. K., & Savale, V. K., 2016. Intellectual Property Rights (IPR).

Shukla, S., Shankar, R., & Singh, S. P., 2014. Food safety regulatory model in India. Food Control, 37: 401-413.

Wambui, J., Karuri, E., Lamuka, P., & Matofari, J., 2017. Good hygiene practices among meat handlers in small and medium enterprise slaughterhouses in Kenya. Food Control, 81: 34-39.

Important links for accessing additional study material

The students can explore the following links for the detailed information on the content provided in the chapter. The videos are adopted from NPTEL-A Project funded by MHRD, Govt. of India to facilitate the students and faculty with E-learning through online Web and Video courses for various streams.

Quality and Safety Aspects of Food

https://www.youtube.com/watch?v=WRYoGiOobqU&t=844s

Quality of Food

https://www.youtube.com/watch?v=Mw1edmzTACQ

GMP/GHP in Food Industry

https://www.youtube.com/watch?v=VeFw_7np8Iw

Food Laws - Why?

https://www.youtube.com/watch?v=yLWRdmT2Fpk

18

Bio-waste Utilization for Value Addition

Ghan Shyam Abrol

Department of Postharvest Technology, College of Horticulture, Rani Lakshmi Bai Central Agricultural University, Jhansi-284 003, Uttar Pradesh

The production of fruit and vegetables has recently increased significantly and even surpassed the total grain production in the country due to increased awareness of consumer and positive government initiatives (Schieber *et al.*, 2001; Sagar *et al.*, 2018; Malenica and Bhat, 2020). But due to limited production and area, the importance of excess production during a particular season and waste management arises at highest priority level. Moreover, fruits and vegetables waste are a serious problem for environment as most of themare being disposed of as landfills or directly disposed off to water bodies such as rivers (Wadhwa and Bakshi, 2013). Fruits and vegetables are perishable in nature and rich source of nutrients and vitamins, which results in microbiological contamination (Coman *et al.*, 2019; Bas-Bellver *et al.*, 2020). Foods wastes are usually organic residues generated by the processing of raw agricultural materials into food and are made up of liquid (wastewater) and solids (Bas-Bellver *et al.*, 2020). The wastewater results from the cleaning processes or in the form of excessive or polluted process water. Its dry material content is typically less than 5% by mass. It possibly also contains organic or inorganic cleaning agents or disinfectants. Solid food wastes with an organic origin have remarkably high water content (mostly about 80% by mass). They are usually characterized by a constant quality and purity due to the forgone processes. The fact that these substances are removed from the production process as undesirable ingredients makes them, wastes.

Agricultural waste, by-products and co-products are usually defined as plant or animal residues that are not (or not further processed into) food or feed, that may even be responsible for additional environmental and economic burdens in the farming and primary processing sectors. In order to further improve resource efficiency and improve agricultural waste management in primary production, it is considered of paramount importance to promote a circular economy approach. Agricultural waste is, mainly, primary residues that can be turned into resources using intensified conversion processes which may yield potentially sustainable bio-products such as fertilisers, energy, materials and molecules. The conversion of this agricultural residue is crucial for supporting the decoupling of economic growth and human well-being from (primary) resources use, preventing putting pressure on land, causing adverse effects on biodiversity and jeopardising global food security.

However, positive progresses and better initiatives has been taken to manage this nutritionally rich wastage in production of by-products. The term "by-product," which is common in industry, points up that these are mostly concealed usable substances, often with a market value. So, the wastes could be considered valuable by-products if there were appropriate technical means to produce products whose value exceeds the cost of reprocessing. Residues in this case cannot be regarded as wastes but become an additional resource to augment existing natural materials. Recycling, reprocessing and eventual utilization of food processing residues offer potential of returning these by-products to beneficial uses rather than their discharge to the environment which cause detrimental environmental effects. There aremany sustainable approaches has been taken out for handling horti-agricultural waste, such as, reuse as soil amendment, animal feed, extracting of bioactive compounds (like starch, pectin etc.), bio-fortification etc. (Plazzotta *et al.*, 2017; Coman *et al.*, 2019; Peng *et al.*, 2019; Salehiyoun *et al.*, 2019; Malenica and Bhat, 2020). This article is to highlight the different methodologies followed to minimise the wastage and its conversion into suitable medium.

Types of Food By-product and Wastes

Five system boundaries have been distinguished in the food supply chains (FSC) which are mainly responsible for generation of wastages.

i) Agricultural production: Wastage of agricultural resources due to environmental conditions, losses due to mechanical damage and/or spillage during harvest operation (e.g. threshing or fruit picking), curing/pre-treatments, on-farm temporary storage and sorting of crops etc.

ii) Postharvest handling and storage: including losses due to spillage and degradation during handling, storage and transportation between farm and distribution.

iii) Processing: including losses due to spillage and degradation during industrial or domestic processing, e.g. juice production, canning and bread baking. Losses may occur when crops are sorted out if not suitable to process or during washing, peeling, slicing and boiling or during process interruptions and accidental spillage.

iv) Distribution: including losses and waste in the market system, at e.g. wholesale markets, supermarkets, retailers and wet markets.

v) Consumption: including losses and waste during consumption at the household level.

Magnitude of By-products/wastes in FoodProcessing

The food processing industry provides vital linkages and synergies between industry and agriculture. Raw materials - grain, livestock, milk, fruit, vegetables etc. are converted into foodstuffs through processing activities which may involve other inputs - energy, water, packaging and so forth. However not all inputs consumed result in useful outputs, and the result is waste.

Table 1: Food processing operations and associated waste

Catagory	Operations	Wastes
Raw material Preparation	Cleaning, Sorting, Grading, Peeling	Cleaning water effluent (BOD or COD), peelings, hair, feathers, grit, blood, contaminated foodstuffs
Size Reduction	Chopping, cutting, slicing, dicing, Milling of Solid foods, pulping Emulsification and homogenization of liquids	Poor quality (too coarse / fine) product with loss of nutritional / sensory characteristics Dust Agglomerates. Waste off-cuts. Fat bearing effluents from colloidal products (e.g. dairy) Risk of pathogenic contamination in emulsification (e.g. dairy)
Mixing and forming	Mixing, forming malformed pieces.	Wrongly proportioned batches, poorly mixed ingredients
Separation and Concentration	Centrifugation, Filtration Expression, Solvent extraction Membrane concentration	Separated solids (e.g. after clarification of liquids press residues (e.g. fruit juice extraction)
Fermentation and use of enzymes	Fermentation, Enzyme technology	Spent biomass.

By-products/wastes Management in fFood Processing

1. Formulation of functional food and source of dietary fibres

Fruit and vegetable are a rich source of nutritious compounds such as polyphenols, dietary fibres, carotenoids, and vitamins, which are beneficial for human health and can be formulated in novel functional food products (Makris *et al.*, 2007; Ajila and Prasada Rao, 2013; Javed *et al.*, 2019). Conversion of residual waste produced during processing into flour found to possess various functional and antioxidant properties. Ferreira *et al.* (2015) has evaluated functional properties of flour obtained from residues generated during beverage manufacture from several fruits and vegetables: orange, watermelon, passion fruit, lettuce, taro, carrot, courgette, mint, spinach, cucumber and rocket. The fruit and vegetable residue flour were used for making of cereal bars and biscuits. Results showed that the obtained flour was suitable for production of functional food. In addition, Černiauskiene *et al.* (2014) have converted pumpkin into flour, while, Juarez-Garcia *et al.* (2006) used unripe banana fruit for developing of banana flour to be used as a source of food enrichment in bakery or other purposes. Bakery based foods prepared from pumpkin flour manifested high content of vitamins, minerals and dietary fiber (Noor Aziah and Komathi, 2009).

Fruit and vegetable-based waste found to have dietary fibres which are found to be preventive of diabetes, cardiovascular diseases, obesity and cancer (Ferguson, 2005; Figuerola *et al.*, 2005; Evans,2019; Koulouris *et al.*, 2019). It has also been converted as pharmaceutical excipient (Domínguez-Robles *et al.*, 2019) and solid oral dosageformulations (Credou and Berthelot, 2014; Banerjee *et al.*, 2017). For production of cellulose, following fruit and vegetable waste is used citrus peels (Güzel and Akpınar, 2019), banana waste (Elanthikkal *et al.*, 2010), potato peel (dos Santos *et al.*, 2012; Sagar *et al.*, 2018), mandarin peel (Hiasa *et al.*, 2014), etc.

2. Proteins and Enzyme Production

Fruit and vegetables wastes especially seeds are very good source of proteins (Banerjee *et al.*, 2017). Certain fruit and vegetable waste have more than 20% of crude protein (Malenica and Bhat, 2020), while different part of fruits and vegetables which is considered as waste also contains good amount of protein for example, potato peels, carrot pomace, apple pomace, citrus peel, green pea peels, mango peel, avocado waste, cabbage leaves, pineapple peel, tomato solid waste etc. (Wadhwa and Bakshi, 2013; Chitturi *et al.*, 2013; Bakshi *et al.*, 2016;Sharma *et al.*, 2016; Sagar *et al.*, 2018). The protein extracted from these types of wastage not only provide the nutrition but also exhibits some health benefits such as protein actinidin extracted from kiwifruit exhibits probiotic activities (Puglisi *et al.*, 2012; Boland, 2013).Beside, excellent source of protein fruits and vegetables are also good source of enzymes like amylases, pectinase, cellulases etc. Enzyme amylases can be extracted from Banana peel, orange peel, potato peels, rice bran etc. and found useful in the production of fruit juices, cheese, chocolate cakes, syrups etc. (Pereira *et al.*, 2017; Oshoma *et al.*, 2019). Invertase enzyme extracted from pineapple peel, pomegranate peel, papaya peel etc. found to be useful in industrial production of artificial sweeteners, chocolates, candies and confectionaries etc. (Sagar *et al.*, 2018; MashettyandBiradar, 2019; Oyedeji *et al.*, 2017; Chelliappan and Madhanasundareswari, 2013). Cellulases extracted from potato peel and orange waste used in various industries such as textile industries and industrial biotechnologies (Imran *et al.*, 2016; Bajaj & Mahjan, 2019; Oyedeji andOjekunle,2018). Xylanases the other enzymes extracted from banana peel, passion fruit peel and orange peel has a wide role in biofuel industry, food industry, textile industry and even used as supplement to animal (Kumar *et al.*, 2018; Singh, 2019, Zehra *et al.*, 2020).

3. Organic Acids and Pigments

Fruits and vegetables waste is a good source of production of organic acids such as, citric and lactic acids which are extensively used in beverage industries, cosmetic and pharmaceutical industries (Table 2). Aspergillus niger and yeastYarrowia lipolyticais generally used for the production of citric acid while, Lactobacillus casei, Lactobacillus delbrueckii, and Lactobacillus plantarum are used for lactic acid production (Imandi *et al.*, 2008; Sagar *et al.*, 2018). In these days naturally extracted pigments has gained importance over synthetic pigments as they exhibits both nutritional and health healing properties. Lycopene extracted from tomato pomace has antioxidant properties and is known to reduce the chances of cancer(Grabowska *et al.*, 2019; Jiang *et al.*, 2019; Kim *et al.*, 2019; Sen, 2019).

Organic acid	Extracted from	References
Citric acid	Kiwifruit peel	Hang *et al.*, 1987
	Apple pomace	Dhillon *et al.*, 2011
	Cacao pod husk	Vriesmann *et al.*,2012
	Pineapple waste	Imandi *et al.*, 2008
Lactic acid	apple peel	Gullon *et al.*, 2008
	mango waste	Jawad *et al.*, 2013
	wheat bran and wheat straw	Panesar and Kaur, 2015
	potato peel, green peasand sweet corn	Ray *et al.*, 2008; Sagar *et al.*, 2018

4. Polyphenols

Today's consumer become more health conscious and including the diet rich in polyphenols in their menu. Polyphenols manifest many important biological health healing and protecting properties such as, scavenging free radicals, prevention of oxidation reactions in food, prevention of oxidative stress etc. (Popa *et al.*, 2008; Sagar *et al.*,2018). Fruits and vegetable waste represent a valuable source of polyphenols. There are various polyphenols hydroxycinnamic acids: ferulic acid, quercetin, ferulic acid hexoside, sinapic acid, myricetin-3-rutinoside,(+)-catechin, epicatechin, gallocatechin, anthocyanins, ellagic acid, gallic acid, punicalagin etc. which are extracted from peel, leaves, husk and pomace of a number of fruits viz. apple, banana, carrot, eggfruit, kiwi, mango, papaya, tomato etc. (Lommen *et al.*, 2000; Saini *et al.*,2019; Waghmare and Kurhade, 2014; Zeyada *et al.*, 2008; Kallel *et al.*, 2014; Akhtar *et al.*, 2015; Kujala *et al.*, 2001).

Conclusion

Fruit and vegetable waste and by-products show highpotential for re-use, recycle and recovery in different forms. A huge amount of waste has been generated every year by different postharvest operation like improper harvesting, sorting, trimming etc. but these waste in different studies has been found very effective in improving the health and utilized in production of different products after certain treatments such as fiber extracted from apple pomace utilized in production of pre- and probiotic rich yogurt. In these days a lot of reach has been carried out for the production of biofuel from these wastes. Pigments produced and the recovery of bioactive compounds through various extraction techniques allows their utilization in food, cosmetics and pharmaceutical industries. But still there is a need to do a lot in this direction as extraction technologies are very costly. Further, an extensive research is required for further standardization, improvement, and development of novel cost-effective sustainable strategies for reducing of agri-food wastes.

References

Ajila, C.M. and Prasada Rao, U.J.S. 2013. Mango peel dietary fibre: Composition and associated bound phenolics. Journal of Functional Foods 5(1): 444–450.

Akhtar, S., Ismail, T., Fraternale, D. andSestili, P. 2015. Pomegranate peel and peel extracts:Chemistry and food features. Food chemistry 174: 417–425.

Bajaj, P. and Mahajan, R. 2019. Cellulase and xylanase synergism in industrial biotechnology. Applied microbiology and biotechnology 103(21–22): 8711–8724.

Bakshi, M.P.S., Wadhwa, M. and Makkar, H.P. 2016. Waste to worth: vegetable wastes as animal feed. CAB Reviews 11(012): 1–26.

Banerjee, J., Singh, R., Vijayaraghavan, R., MacFarlane, D., Patti, A.F. and Arora, A. 2017. Bioactives from fruit processing wastes: Green approaches to valuable chemicals. Food Chemistry 225: 10–22.

Bas-Bellver, C., Barrera, C., Betoret, N. & Seguí, L. 2020. Turning Agri-Food Cooperative Vegetable Residues into Functional Powdered Ingredients for the Food Industry. Sustainability 12(4): 1284.

Boland, M. 2013. Kiwifruit proteins and enzymes: Actinidin and other significant proteins. Advances in Food and Nutrition Research 68: 59–80.

Chelliappan, B. and Madhanasundareswari, K. 2013. Production and optimization of growth conditions for invertase enzyme by aspergillus sp., in solid state fermentation (SSF) using papaya peel as substrate. The Journal of Microbiology, Biotechnology and Food Sciences 3(3): 266–269.

Chitturi, S., Talatam, V.G. and Vuppu, S. 2013. Studies on protein content, protease activity, antioxidants potential, melanin composition, glucosinolate and pectin constitution with brief statistical analysis in some medicinally significant fruit peels. Der Pharmacia Lettre 5(1): 13–23.

Coman, V., Teleky, B.E., Mitrea, L., Martău, G.A., Szabo, K., Călinoiu, L.F. &Vodnar, D.C. 2019. Bioactive potential of fruit and vegetable wastes. Advances in Food and Nutrition Research 91: 157–225.

Credou, J. and Berthelot, T. 2014. Cellulose: from biocompatible to bioactive material. Journal of Materials Chemistry B, 2(30): 4767–4788.

Dhillon, G.S., Brar, S.K., Verma, M. and Tyagi, R.D. 2011. Enhanced solid-state citric acid bioproduct using apple pomace waste through surface response methodology. Journal of applied microbiology 11(4): 1045–1055.

Domínguez–Robles, J., Stewart, S.A., Rendl, A., González, Z., Donnelly, R.F. and Larrañeta, E. 2019. Lignin and cellulose blends as pharmaceutical excipient for tablet manufacturing via direct compression. Biomolecules 9(9): 423. https://doi.org/10.3390/biom9090423

Elanthikkal, S., Gopalakrishnapanicker, U., Varghese, S. and Guthrie, J.T. 2010. Cellulose microfibers produced from banana plant wastes: Isolation and characterization. Carbohydrate polymers 80(3): 852–859.

Evans, C.E.L. 2019. Dietary fibre and cardiovascular health: a review of current evidence and policy. Proceedings of the Nutrition Society 79(1): 61–67.

Ferguson, L.R. 2005. Does a diet rich in dietary fibre really reduce the risk of colon cancer?. Digestive and Liver Disease 37(3): 139–141.

Ferreira, M.S., Santos, M.C., Moro, T.M., Basto, G.J., Andrade, R.M. and Gonçalves, É.C. 2015. Formulation and characterization of functional foods based on fruit and vegetable residue flour. Journal of Food Science and Technology 52(2): 822–830.

Grabowska, M., Wawrzyniak, D., Rolle, K., Chomczyński, P., Oziewicz, S., Jurga, S. and Barciszewski, J. 2019. Let food be your medicine: nutraceutical properties of lycopene. Food & function 10(6): 3090–3102.

Gullon, B., Yanez, R., Alonso, J.L. andParajo, J.C. 2008. L-Lactic acid production from applepomace by sequential hydrolysis and fermentation. Bioresource Technology 99(2): 308–319.

Güzel, M. and Akpınar, Ö. 2019. Production and characterization of bacterial cellulose from citrus peels. Waste and Biomass Valorization 10(8): 2165–2175.

Hang, Y.D., Luh, B.S. andWoodams, E.E. 1987. Microbial production of citric acid by solid statefermentation of kiwifruit peel. Journal of Food Science 52(1): 226–227.

Imandi, S.B., Bandaru, V.V.R., Somalanka, S.R., Bandaru, S.R. andGarapati, H.R. 2008. Applicationof statistical experimental designs for the optimization of medium constituents for theproduction of citric acid from pineapple waste. Bioresource Technology 99(10): 4445–4450.

Imran, M., Anwar, Z., Irshad, M., Asad, M.J. and Ashfaq, H. 2016. Cellulase production from species of fungi and bacteria from agricultural wastes and its utilization in industry: a review. Advances in Enzyme Research 4(2): 44–55.

Javed, I.M., Waseem, A.M. and Ammad, R. 2019. Vegetables as a source of important nutrients and bioactive compounds: their human health benefits. MOJ Food Process Technol. 7(4): 136–146.

Jawad, A.H., Alkarkhi, A.F., Jason, O.C., Easa, A.M. andNorulaini, N.N. 2013. Production of thelactic acid from mango peel waste–Factorial experiment. Journal of King Saud University-Science 25(1): 39–45.

Jiang, L.N., Liu, Y.B. and Li, B.H. 2019. Lycopene exerts anti-inflammatory effect to inhibitprostate cancer progression. Asian Journal of Andrology 21(1): 80.

Juarez-Garcia, E., Agama-Acevedo, E., Sáyago-Ayerdi, S.G., Rodriguez-Ambriz, S.L. and BelloPerez, L.A. 2006. Composition, digestibility and application in breadmaking of banana flour. Plant foods for human nutrition 61(3): 131. doi: 10.1007/s11130-006-0020-x

Kallel, F., Driss, D., Chaari, F., Belghith, L., Bouaziz, F., Ghorbel, R. andChaabouni, S. E. 2014.Garlic (Allium sativum L.) husk waste as a potential source of phenolic compounds:Influence of extracting solvents on its antimicrobial and antioxidant properties. IndustrialCrops and Products 62: 34–41.

Koulouris, A.I., Luben, R., Banim, P. and Hart, A.R. 2019. Dietary Fiber and the Risk of Pancreatic Cancer. Pancreas 48(1): 121–125.

Kujala, T., Loponen, J. andPihlaja, K. 2001. Betalains and phenolics in red beetroot (Beta vulgaris)peel extracts: extraction and characterisation. ZeitschriftfürNaturforschung C 56(5–6): 343–348.

Kumar, V., Dangi, A.K. and Shukla, P. 2018. Engineering thermostable microbial xylanases toward its industrial applications. Molecular biotechnology 60(3): 226–235.

Lommen, A., Godejohann, M., Venema, D.P., Hollman, P.C.H. and Spraul, M. 2000. Applicationof directly coupled HPLC− NMR− MS to the identification and confirmation of quercetinglycosides and phloretin glycosides in apple peel. Analytical chemistry 72(8): 1793–1797.

Makris, D.P., Boskou, G. and Andrikopoulos, N.K. 2007. Polyphenolic content and in vitro antioxidant characteristics of wine industry and other agri-food solid waste extracts. Journal of Food Composition and Analysis 20(6): 125–132.

Malenica D. and Bhat R. 2020. Review article: Current research trends in fruit and vegetables wastes and by-products management-Scope and opportunities in the Estonian context. Agronomy Research 18(S3): 1760–1795, 2020 https://doi.org/10.15159/AR.20.086

Mashetty, S.B. and Biradar, V. 2019. Orange Peel as Novel Substrate for Enhanced Invertase Production by A. niger in Solid State Fermentation. Int. J. Curr. Microbiol. App. Sci. 8(4): 1114–1121.

Noor Aziah, A.A. and Komathi, C.A. 2009. Acceptability attributes of crackers made from different types of composite flour. International food research journal 16(4): 479–482.

Oshoma, C.E., Eguakun-Owie, S.O. and Obuekwe, I.S. 2019. Utilization of Banana Peel as a Substrate for Single Cell Protein and Amylase Production by Aspergillus niger. African Scientist 18(3): 143–150.

Oyedeji, O., Bakare, M.K., Adewale, I.O., Olutiola, P.O. and Omoboye, O.O. 2017. Optimized production and characterization of thermostable invertase from Aspergillus niger IBK1, using pineapple peel as alternate substrate. Biocatalysis and Agricultural Biotechnology, 9: 218–223.

Oyedeji, O. and Ojekunle, O.O. 2018. Cellulase Production by Penicillium citrinum using Brewers Spent Grain and Pineapple Peels as Cheap, Alternate Substrates. International Journal of Sciences 7(01): 74–83.

Panesar, P.S. and Kaur, S. 2015. Bioutilisation of agro-industrial waste for lactic acidproduction. International journal of food science & technology 50(10): 2143–2151.

Peng, W., Ma, Q., Wang, Z. and Xie, Z. 2019. Research Progress On Comprehensive Utilization of Fruit and Vegetable Waste. In E3S Web of Conferences (Vol. 131, p. 01106). EDP Sciences.

Pereira, C.R., Resende, J.T.V., Guerra, E.P., Lima, V.A., Martins, M.D. and Knob, A. 2017. Enzymatic conversion of sweet potato granular starch into fermentable sugars: feasibility of sweet potato peel as alternative substrate for α-amylase production. Biocatalysis and agricultural biotechnology 11: 231–238.

Plazzotta, S., Manzocco, L. and Nicoli, M. C. 2017. Fruit and vegetable waste management and the challenge of fresh-cut salad. Trends in food science & technology 63: 51–59.

Popa, V.I., Dumitru, M., Volf, I. andAnghel, N. 2008. Lignin and polyphenols asallelochemicals. Industrial crops and products 27(2): 144–149.

Puglisi, I., Petrone, G., Piero, A. and Lo, R. 2012. Role of actinidin in the hydrolysis of the cream milk proteins. Food and Bioproducts Processing 90(3): 449–452.

Ray, R.C., Mohapatra, S., Panda, S. and Kar, S. 2008. Solid substrate fermentation of cassava fibrousresidue for production of a-amylase, lactic acid and ethanol. J Environ Biol. 29(1/2): 111–115.

Sagar, N.A., Pareek, S., Sharma, S., Yahia, E.M. & Lobo, M.G. 2018. Fruit and vegetable waste: Bioactive compounds, their extraction, and possible utilization. Comprehensive Reviews in Food Science and Food Safety 17(3): 512–531.

Saini, A., Panesar, P.S. andBera, M.B. 2019. Valorization of fruits and vegetables waste throughgreen extraction of bioactive compounds and their nanoemulsions-based deliverysystem. Bioresources and Bioprocessing 6(1): 26. https://doi.org/10.1186/s40643-019-0261-9

Salehiyoun, A.R., Sharifi, M., Di Maria, F., Zilouei, H. & Aghbashlo, M. 2019. Effect of substituting organic fraction of municipal solid waste with fruit and vegetable wastes on anaerobic digestion. Journal of Material Cycles and Waste Management 21(6): 1321–1331.

Schieber, A., Stintzing, F.C.and Carle, R. 2001. By-products of plant food processing as a source of functional compounds—recent developments. Trends in Food Science and Technology 12(11): 401–413

Sen, S. 2019. The Chemistry and Biology of Lycopene: Antioxidant for HumanHealth. International Journal of Advancement in Life Sciences Research 2(4): 8–14.

Sharma, R., Oberoi, H.S. and Dhillon, G.S. 2016. Fruit and vegetable processing waste: renewable feed stocks for enzyme production. In: Dhillon, G.S., Kaur, S., editors. Agro-industrial wastes as feedstock for enzyme production: apply and exploit the emerging and valuable use options of waste biomass., Academic Press Elsevier, London, UK. pp. 23–59.

Singh, B. 2019. Production, characteristics, and biotechnological applications of microbial xylanases. Applied Microbiology and Biotechnology 103(21–22): 8763–8784.

Vriesmann, L.C., Teófilo, R.F. and de Oliveira Petkowicz, C.L. 2012. Extraction and characterizationof pectin from cacao pod husks (Theobroma cacao L.) with citric acid. LWT 49(1): 108–116.

Wadhwa, M. and Bakshi, M.P.S. 2013. Utilization of fruit and vegetable wastes as livestock feed and as substrates for generation of other value-added products. Rap Publication 4: 1-67.

Waghmare, J.S. and Kurhade, A.H. 2014. GC-MS analysis of bioactive components from bananapeel (Musa sapientum peel). European Journal of Experimental Biology 4(5): 10-15.

Zehra, M., Syed, M.N. & SOHAIL, M. 2020. Banana Peels: A Promising Substrate for the Coproduction of Pectinase and Xylanase from Aspergillus fumigatus MS16. Polish Journal of Microbiology 69(1): 19–26.

Zeyada, N.N., Zeitoum, M.A.M. and Barbary, O.M. 2008. Utilization of some vegetables and fruitwaste as natural antioxidants. Alexandria Journal Food Science & Technology 5: 1–11.

Černiauskiene, J., Kulaitiene, J., Danilcenko, H., Jariene, E. and Jukneviciene, E. 2014. Pumpkin fruitflourasasourceforfoodenrichmentindietaryfiber. NotulaeBotanicaeHortiAgrobotanici Cluj-Napoca 42(1): 19–23.

19

Processing and Value Addition of Banana and Its By-products

Suresh Kumar P., Amelia K.D., Divya P. and Uma S.

ICAR-National Research Centre for Banana, Tiruchirappalli-620102, Tamil Nadu

As the consumers nowadays demand convenient food products with high nutrition, the nutritional abundance in banana can be utilized for developing such products. World production of banana is estimated at 155 million MT of which, about 68% was classified as bananas and 32% as plantains (FAO, 2019). In India, banana as an enterprise is generating around US $7 billion and providing livelihood to millions of farmers. India contributes 20.48% world banana production with the production of 31.75 million tons in an area of 8.98 lakh hectares (NHB, 2020). Banana alone contributes 2.3% to the Agricultural GDP.Presently, India shares 0.34% of total export of banana with the value of over 600 crores and the export volume of 146.7 thousand metric tons (APEDA, 2020).

Nonetheless of agriculture sharing major chunk of labour force of over 50% with 400 million dependent populations, the processing industry gives employment to less than 2 million populations with the share of only 0.5%. Due to mishandling of the produce, about 25-40% is being wasted and only 2% is processed into value added products, the remaining being used in the raw form. This leads to price imbalance and large price variations both spatial and temporal placing farmers in a disadvantage situation. This loss in value could be easily minimized with the scientific handling of the perishable commodities and by establishing permanent rural-cluster based agro-processing plants instead of having plug-in and plug-out kind of processing industries, which mushrooms only when there is a crisis.India is hardly processing banana into a value-added product due to its high domestic demand. However, the need for processing arises with the change in consumer behaviour and food consumption pattern. According to an INIBAP survey, about 5% of total world production of dessert banana and 24% of plantain is processed.

After harvesting the fruit bunch, huge plant bio-waste is generated, out of which pseudostem (30-34%), flower & bracts (5%) and rhizome (12-14%) together contribute 50 per cent of the banana plant biomass. Currently less than 2% of these wastes production is used for human consumption and for production of fibre, the remaining are incinerated and wasted. In indigenous systems of medicine, pseudostem juice is a well-known remedy for urinary disorders, stomach troubles like diarrhoea, dysentery and flatulence. The usage of banana fibre for textile purposes predates written history. The evidence of this can be found in epics like Ramayana where Sita and Rama wore "Naravastra" clothes made from banana fibres. Apart from India, past history and references reveal that the banana fibre cloth was earlier made around 13th century in Japan. Japanese currency, the Yen, is made out of banana fibre. The banana fibres can be utilized to produce yarns of different fineness to make textile products with different qualities for specific end use. In olden days extracted banana fibre was used for making rugs, ropes and for tying flowers. Philliphines is having a banana species Abaca (*Musa textilis*, Manila Hemp) specially meant for the extraction of fibre. The research for the use of natural fibres has seen a tremendous growth because of their low cost, easy availability, excellent mechanical properties, high specific strength, Non-abrasiveness, eco-friendly and biodegradable characteristics.

Nutritional Values of Banana

Bananas are cheap energy source which contains 27% carbohydrate, 1.0% protein and 0.3% fat providing 116 kcal of energy per 100 g of flesh (Uma *et al.*,2008). It has a plenty of micro and macro nutrients. Banana is highly nutritious fruit and pulp (edible portion) of ripe fruit is rich in vitamin A, B-complex (thiamine, riboflavin, niacin, pentothenic acid, pyridoxine and folic acid), ascorbic acid vitamin E (Someya *et al.*, 2002; Amorim *et al.*, 2009) and micro (iron, copper and zinc) and macro minerals (potassium, calcium, phosphorus and magnesium)and contains health-beneficial ingredients such as resistant starch (>22%), dietary fibres, rapidly digestible starch and slowly digestible starch. In addition, fructooligosaccharides and polyphenols (catechin, epicatechin, epigallocatechin and gallic acid) are other ingredients present in unripe banana. Flower also a good source of dietary fibre, vitamin A, vitamin B2 and minerals such as potassium and calcium and central core stem too possess dietary fibre and vital nutrients. Hypertension, constipation, depression, hangovers, heart burns, morning sickness, premenstrual syndrome, seasonal affective disorder, ulcer, stroke, stress and geriatric problems are some of the disorders known to be managed by consumption of bananas.

Table 1. Nutritional values of banana pulp, peel, flower and pseudostem

Composition	Plantain Pulp	Banana Pulp	Peel	Flower	Central core stem
Moisture (%)	64.1-67.6	70.2 - 73.8	6.7	--	--
TSS (°Brix)	3.5-5.1	19.2	--	--	--
Total carbohydrate (%)	32.7	25	--	--	--
Protein (%)	1.0-1.34	1.25	1.3	2.2	0.35
Fat (%)	0.36-0.39	0.5	1.7	0.4	0.12
Glucose (%)	--	2.0	1.4	--	--
Fructose (%)	--	3.5	3.2	--	--
Sucrose (%)	--	12.0	2.6	--	--
Starch (%)	24.3	9.0	2.2	4.0	2.2
Crude fibre (%)	4.8	9.0	3.4	--	2.0
Total dietary fibre (%)	8.0	7.7	12	6.0	4.0
β-Carotene (μg/100 g)	1000	75 - 1000	--	100	--
Thiamine (mg/100 g)	0.05	0.04	--	20	--
Riboflavin (μg/100 g)	0.05	0.07	--	120	--
Niacin (mg/100 g)	0.7	0.6	--	--	--
Pantothenic acid (mg/100 g)	0.26	0.40	--	--	--
Pyridoxine (mg/100 g)	--	0.40	--	--	0.24
Folic acid (μg /100 g)	23.0	25.0	--	--	--
Vitamin C (mg /100 g)	18	20	6.0	--	--
α-Tocopherol (mg/100 g)	--	0.2	0.10	--	--
Phylloquinone (μg/100 g)	--	0.5	--	--	--
Potassium (mg/100 g)	500	450	280	450	240
Phosphorous (mg/100 g)	34	25	--	70	35
Calcium (mg/100 g)	15	10-140	20	55	30
Magnesium (mg /100 g)	30	40	--	40	19
Sodium (mg /100 g)	4	1	2.4	24	24
Iron (mg /100 g)	0.6	0.4	0.5	35	30
Manganese (mg /100 g)	--	0.30	0.7	2.0	0.08
Zinc (mg /100 g)	0.14	0.15	--	--	12
Copper (mg /100 g)	--	0.08	--	13.0	4.0

Amorim *et al.* (2011)

Value Added Products from Banana

Value added products from banana are made from banana of raw and ripe bananas, based on the preferable end product. Generally the value added products from banana can be grouped into 3 major categories.

- Raw banana based products
- Ripe banana bases products
- Products from other parts of banana plant

A gist of available value added products from banana is given in Fig.2

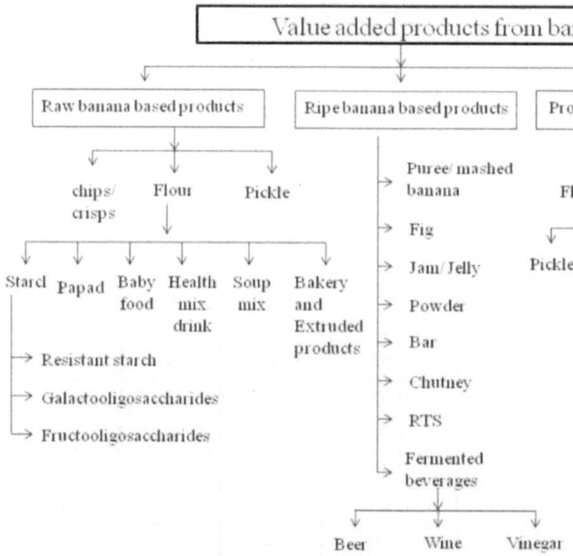

Fig. 1. Value added products from banana

Raw Banana Based Products

Banana chips/crisps

Banana chips/crisp is the major processed products of raw banana, which can be seen in the market shelf of south India, Thailand and Philippines. Because of the growth in snack foods the processing of bananas into chips has the potential to become an important industry in countries where bananas are produced. Banana chips are produced by deep frying of unripe banana slices till they become crisp. Hydrocolloids (alginate, CMC and pectin) has the capability to reduce the oil absorption in fried banana chips.

Banana flour

New economical strategies are now considered for banana use, such as the production of banana flour when the fruit is unripe. Banana flour prepared from matured unripe banana with high starch content (available and resistant) can be blended with other cereal flours and can be used as a potential ingredient for bakery products containing slowly digestible carbohydratesexample for preparing bread, cookies, biscuits, baby food, soup mix, extruded products (noodles, pasta, and breakfast cereal), health mix and unleavened chapattis and *rotis*. Supplementation of banana flour to wheat flour aided in increasing resistant starch and phenolic content of wheat breads and the glycemic index for the banana flour bread was 65.08%, which was significantly lower thancontrol bread (81.88%), suggesting a "slow carbohydrate" feature for the banana flour based products.

Starch and Modified Starches

As classified by Brown *et al.* and Englyst *et al.* (1992), RS is classified into four groups, namely RS1; physically inaccessible (partly milled grains and seeds) RS2; resistant granules (Raw potato, green banana, some legumes, and high-amylose starches) RS3; retrograded (Cooked and cooled potato, bread, and cornflakes) RS4; chemically modified (Etherized, esterified, or cross-bonded starches). Resistant starch has physiological functions similar to those of dietary fibre, are digested very slowly and incompletely in the small intestine. Green banana flour contains high amounts of resistant starch. This predominant characteristic has attracted researchers to develop processed food products and to evaluate in-vitro studies during starch digestion.

Extruded Products

Unripe banana flour (UBF) is an excellent source to make extruded products (cold extrusion) of high quality. Pasta products containing banana flour exhibit a low rate of carbohydrate enzymatic hydrolysis and they could help broaden the range of low-glycaemia index foods available to the consumer.Moreover, this food contains increased levels of indigestible compounds (resistant starch and non-starch polysaccharides) that may be beneficial for intestinal health

Low Glycemic, Fibre Rich Prebiotic Products from Raw Banana

Due its higher resistant starch content banana could be used for making different functional products with prebiotic characteristics which could be made available to even diabetic patients. Extruded snacks like pasta, noodles and vermicelli can be prepared by partial replacement of traditional *maida* which has more gluten with banana flour or modified starches. Banana bread, muffin could be produced with the combination of banana flour, peel powder and addition of low glycemic starches from Banana.

Weaning Food

A combination of malted maize and soybean, roasted groundnut and cooking banana in the ratio of 50:15:15:20 gave a very recommendable weaning food for infants between the ages of 6 months and 2 years. Some products like health mix drink, Soup mix, Banana sauce, and pickle are commercially available in local markets of banana producing region and these paves way to investigate it health benefits and standard protocol for commercialization

Ripe Banana Based Products

Ripe banana fruit is used for treating gastric problem, since it is near to alkaline in nature, neutralizes the acid conditions and helps in formation of thick protective mucus layer. Ripe bananas are excellent food for high energy required people (athletes, manual workers), because of its high sugar and potassium content (4.10-5.55 mg per 100g dry weight) and magnesium which are highly require for muscle performance.

Dehydrate Ripe Banana (Banana fig)

Banana figs are dehydrated ripe banana, traditionally popular among dried fruits. It is cheapest among all existing dry fruits in market.Ecuador is the leading producer of fig in the world. Varieties with high TSS (>22 brix), like udhayam and karpooravalli are suitable for making figs. Sun/ hot air dried banana figs are available in the shelves of local markets. For export market, quality and certification are highly essential.

Banana puree/ Mashed banana

Fruit puree is a very complex system consisting basically of a dispersing medium (serum) and suspended particles (pulp). The serum contains low (mainly sugars, salts and organic acids) and high molecular weight solutes, the removal of the latter leaving a Newtonian solution of low molecular weight species. The possible cause may be soluble pectic substances which generally impart to the serum a non-Newtonian behaviour, yield stress and time dependent properties, and the extent of these properties depends on the puree concentration. Aseptic processing is widely adopted method for banana puree preparation. Banana peel and pulp have been shown to contain biogenic amines, such as serotonin, dopamine and nor epinephrine.Production purees may be a good way for the utilization of surplus stocks, as they serve as a base material for preparation of ice cream, baby foods, bakery products, jam, jelly, RTS, sip ups, chutney, fruit bar, fermented beverages etc.

Jam/Jelly, Ready to serve (RTS) and Others

Jam and jelly from banana have been commercialized before World WarII and it is a well known fact that It banana is a part of mixed fruit jam. Clarified banana juice and juice based ready to drink products, sauce, squash, sweet chutney, suspended juices could be prepared using ripe banana.

Innovative Products From Banana Waste

Banana Wine

If alcohol production per hectare is calculated,bananas are amongst crops giving the highest yield (93-104L).The production of industrial or medicinal alcohol (ethanol) from bananas has been carried out commercially for many years in a variety of countries. Banana wine is produced by fermenting the enzyme treated clear banana juice with wine yeast viz .*Saccharomyces cerevisae var. ellipsoides*. The fermentation is carried out for about 3 weeks followed by filtration, clarification, and bottling. The pasteurized wine is stored in bottles for aging. The alcohol content of banana varies from 9-12%.The word vinegar meaning 'sour wine'has been in use in the English language since the 14th century. Vinegar has been used as a seasoning in cooking since ancient times. Consumers are now beginning to appreciate the health benefits of drinking vinegar and hence there is a growing demand for fruit vinegar products that are sold as a health food.

Edible Banana Flowers

Banana flowers are healthy and edible. Just like the stem, flowers have so much fibre in them. Banana flowers are rich in fibres and antioxidants. Bioactive compounds can be extracted from the flowers which can be used in cosmetic and pharmaceutical industries. Minimal processing of banana flowers will extend it reach to the urban and semi urban markets. NRCB has developed technologies for minimal processing of flowers to attract the urban and semi urban markets. A *thokku* produced using banana flower is rich in nutrients and packed with nutraceutical properties.

Bananas Peel Powder

With its high calcium, banana peel powder is a healthy powder. It is used in food, pharmaceutical and medical drugs. The quality of banana powder is determined by the colour, flavour, texture and moisture content. These are affected by the varieties of bananas and processing operations specially blanching process. Good quality powder is produced from the bananas of right variety and degree of ripeness.

Banana Peel Based Muffins

Banana peel is rich in dietary fiber, protein, essential amino acids, polyunsaturated fatty acids and potassium. They possess good antioxidant compounds like polyphenols, carotenoids and catecholamines. The banana peel flour could be valuable for developing new functional foods with improved nutritional value and textural qualities for health concerned consumers. Nendran peel powder used for making muffins for replacement of refined wheat flour to free from gluten.

Source of Starch, Pectin, and Cellulose

Starch, pectin, and cellulose are used in the food industry as gelling agent, thickening agent and stabilizers. Starch, a group of carbohydrates, is commercially available and produced from plants such as corn, potato, rice, wheat, and cassava. Banana by-products that can be processed into edible starches includes the pith of pseudostem and the green culled banana which are removed during fruit selection and processing. Similarly industrial products like pectin and cellulose can be extracted from the lingo-cellulosic wastes of banana.

Peel and Other Waste as a Source of Xylo-oligosaccharides & Vinegar

Xylo-oligosaccharides (XOS), a Prebiotic, is the only nutraceutical that can be produced from lignocellulosic biomass. Production of XOS from agricultural residues offers great scope to the nutraceutical industries as the raw material is cheap and abundantly available. The major advantages of XOS consumption, apart from elective growth stimulation of beneficial gutmicroflora, include reduction of blood glucose and cholesterol, reduced pro-carcinogenic enzymes in gastrointestinal tract, enhanced mineral absorption from large intestine and immune-stimulation. The sweet taste of XOS enables it to be used as an artificial sweetener. Similarly, vinegar could be produced from the banana wastes with the solid fermentation process.

Source of Dietary Nutrients

By-products of banana and plantain have nutrient content as a potential source of dietary food components such as carbohydrate, proteins, dietary fibres, and minerals for human consumption. Banana pith from the pseudostem has long been eaten as vegetables. It contains considerable amount of starch, sugars and minerals. In most of the southeast regions banana inflorescence is consumed as vegetables and salad for a very long time.

Natural bio-colorant

The anthocyanins, a subclass of flavonoids are an important pigment group that is responsible for the red, purple, and violet colours of the banana inflorescence. Anthocyanin's are considered to be a good bio colorant due to its attractive colours, moderately stable in food systems, water-solubleand proven to be beneficial to health. The abundance of anthocyanin's reported in the banana inflorescence bracts, ranging from 14 – 32 mg anthocyanin/100 g bracts, mainly comprising of cyanidin-3-rutinoside compound, could potentially be exploited as a cheap source of natural food colorant. The content is slightly higher than the commercially available anthocyanin's from red cabbage. Natural bio colorant such as the anthocyanin's remains in demand not only due to its health promoting properties but also the increase in demand on natural foods.

Center Core Stem Based Designer Cookies

Centre core stem powder exhibited higher amount of antioxidant compounds. Nendran center core stem powder incorporated cookies have immense potential for food industry to develop functional foods due to their high nutrient content. They had higher free radical scavenging activity of 0.69mg/100g.

Banana Rhizome

It is estimated that 5-7 tons of rizome can be extracted from 2500 banana plant. Banana rhizome is used as a food and has many medicinal properties as well and it can be stored safely up to 3 months without any spoilage. In Southern parts of India, it is believed that it helps to detoxify the body by functioning as a diuretic. Banana rhizome pickle, juice and juice based products like jelly and candy, soup mix and ice cream mix are some products that can be made using banana rhizome.

Utility Products from Banana Waste

Extraction Methods for Banana Fibre

Banana fibre can be extracted by different methods viz. chemical, mechanical and biological by which heavily coated, non□cellulosic gummy material from the cellulosic part of plant fibres get removed and render them clean and spinnable. Manual extraction needs loads of time and waste of human resource. Mechanical extraction uses Raspador machine. In this machine, fibre extraction is done by mechanical extraction process, which is eco-friendly way to get fibre of both good quality and quantity. In this process the fibre is extracted by inserting the pseudo stem sheaths one by one into a machine. Water and non-fibrous elements are removed from the bottom of the machine. Extracted banana fibres are being exported to overseas market. During the extraction process, substantial amount of wastes are generated, which are mainly non-fibrous element. Machine extracted fibres are low in quality and cheaper in cost when compared to manual one.

Banana Pseudostem

Sheaths from Pseudostem

Raspador machine for fibre extraction

Shade drying of extracted fibre

Softening of Banana Fibre

Fibres undergo degumming to reduce the plants' glue-like gum content and render them clean and spinnable. Since, banana pseudostem has substantial pectin, lignin, cellulose, hemicellulose etc as cell wall component and their removal is crucial for quality fibre extraction. Due to the presence of adherent pith in the fibre produces inferior quality of the product. Fibres in biological material exists in a complex structure of ligno-cellulose that are not easily separated during fibre extraction thus, degumming is required to dissolve them during fibre extraction process. There are different methods of degumming like chemical, mechanical and biological for removal of heavily coated, non-cellulosic gummy material from the cellulosic part of plant fibres. The chemical method uses alkali for the removal of fibres. Alkali treatment has been proven effective to remove the impurities from the fibre, decreasing moisture adsorption and enabling mechanical bonding but due to disposal problem it creates environmental pollution. Mechanical methods require high energy and fail to remove the gummy material sufficiently from the fibre bundle surface. In the processing of textile fibres, enzymes can be used as eco-friendly alternatives to chemical processes. Enzymatic hydrolysis caused an increase in the elastic modulus and tensile strength. The degumming with microbial enzymatic process has been proven to reduce the consumption of chemicals and energy in different fibre crops. The enzymatic degumming serves as a good alternative to reduce pollution and cause less fibre damage. Pectinolytic enzymes can be used to the degumming of natural fibres of textile importance. Pectinases are a group of enzymes which can breakdown the pectic materials. The microbial consortium uses dried banana pseudostem as substrate for pectinase production by fermentation. ICAR- NRCB has developed technologies combining laccase and pectinase enzymes to soften the fibres. Biological procedures yield more fibre bundles than other two procedures without any harm to the environment. But the usage of enzymes for common man is still a riddle.

Application of Banana Fibre and Its Products

Textiles

Currently, most of the fibers from banana are used for handicraft products, textiles. The demand for textiles and ready-made garments is increasing in India with an increasing population and spending power. Relatively higher tensile strength and stiffness of banana fibre make it promising fibre material. Longer fibre of banana results in more yarn production. Moreover, the higher yarn strength of banana fibre facilitates for blending with other natural or synthetic fibres for production of blended fabric and textiles. Banana fibers can be blended with jute and cotton to make banana blended fabrics. Banana-fibre yarns make the cloth brighter, impart better dyeing properties and can also be bleached, for good commercialization a steady bulk supply is necessary.

Handricafts

Banana fiber and Scuture waste (obtained from the extraction of banana fiber) has been used for making paper and paper-based products. The lignin content has been removed by a chemical process.Handmade paper and paper products can be made.

Natural Adsorbent

Banana fibre can act as a natural adsorbent. The fibre has high porosity and natural capillary action making it better in action. Banana fibers can be used in sorbent socks, pillow and booms or as a loose fibre to clean up spills. The fibre is considered to be good on account of its durability, strength and resistance to changes in temperature and moisture. It is considered as agricultural waste material, though those fiber exhibits good tendency of absorbent characteristics. This can be an alternative material for usage of synthetic fiber producing non-biodegradable napkins as well; it can be a good cost effective, eco-friendly and sustainable end product.

Reinforced Composites

Natural fiber-reinforced composites have attracted the attention because they are turning out to be an alternative solution to the ever depleting petroleum sources.Banana fibres are potential engineering materials having considerable scope to be used as an environmental friendly reinforcing element for manufacturing of polymer based green materials.They are superior in quality in terms of strength, fineness and spinabilty compared to bamboo and ramie fibre which make it a esirable source for natural fiber-reinforced composites. Banana fibre was found to be an effective reinforcement in a polymer matrix for making cost effective composites and several works have been carried out to examine its competitiveness as a building material.In countries where there is a lack of softwood fibre, banana fibre is a viable alternative fibre resource for commercial consideration.

Bioremediation Agent

Banana fibers can be used as a natural material for bacteria to thrive that can be used in bioremediation. Banana fibres provide the medium and can be used in remediation industry as a natural bioremediation agent.

Water Purifier

Banana fibre can be used as a filtration agent in waste water treatment industry. The natural affinity of banana fibre to oils and organics and its tendency to repel water, makes it a natural alternative for filtration agent for industrial and municipal waste treatment. Waste water is often contaminated with oils and other organic materials.

Pulp Industry

Banana pseudostem fibre can be used in pulp industry to produce papers. Whitened banana Kraft cellulose pulp can be used to make sheets whose characteristics are similar to japan paper. In Costa Rica, industrial production of paper from banana pseudostem is going on and a whole range of products labeled “Earth” which includes paper, cardboard, note pads, letter paper, envelopes, post cards and note books have been launched. The banana fibre is reported to be of high strength and it is used to make tea bags and currency notes. In Germany, work is in progress to develop banana fibre lining for car interiors.

Cellulose Fibers

Fibers from different source of banana namely, sheath, peduncle, pseudostem & leaf midrid are mainly composed of cellulose, hemicellulose, lignin, wax and pectin. Among those content, cellulose comprises about one-third in the plant cell wall and their properties may vary based on the type of cellulose present in the source. Natural cellulose based fibers originated from bio resources had enough potentials and it is in need of new composite formulation in order to enhance their mechanical properties. Cellulose fibers has been used as a reinforcement with polypropylene binders for manufacturing of bioplastics and other significant application like extruded composites, derivatives products with cellophane, combination with diatomaceous earth or other filtration media, absorbent material as a stationary phase in thin layer chromatography. From the pseudo stem of banana good quality fibers can be extracted along with other plant components with bioconversion potential.

Bio-ethanol Production

Most of the banana waste utilized as feedstock and organic manure sold at a low price. Since banana peel possess low lignin content, which acts as an alternative source for the production of biofuel especially bioethanol. Banana waste generally discarded or dumped in land that causes environmental problems; thereby initiates the growth of harmful microorganisms to produce toxins. Decomposition of waste was undertaken to explore the possibility of converting different banana waste into renewable energy like bioethanol production.

Ferulic Acid

Agricultural lignocellulosic residues are the major potential source of low-cost raw material to commercialize the ferulic acid production. The abundance of this compound in pseudostem of banana and agricultural byproducts combined with the vast array of known microbial transformations of these compounds make them attractive for biotransformation research.

Disposable Plates

Banana leaves have a wide range of applications because they are large, flexible, waterproof and decorative. Banana leaf plates are best described as a revolutionary disposable cutlery and containers are products that are a part of our day to day life. Banana leaf plate making is a state of the art to develop biodegradable and compostable alternatives to petrochemical based plastics and polystyrene. These plates can be easily disposed and are environment friendly.

Bioplastic

The food processing plants used to dispose of banana peel in a typical landfill. The conversion of banana peel into a useful material would thus bring an additional economic gain for the agricultural industry. One of the fruit peel have many advantages like higher biodegradability, renewability and good oxygen barriermost common waste forms of starch in the banana peels. Biopolymers developed from properties which makes them the most suitable alternative for some commercial applicationswith the above advantages as the primary requirements of a package.

Animal Feed

Agricultural and food processing wastes such as rice bran, corncobs, and pineapple waste and sugarcane bagasse have been utilized as an alternative source of animal feed production which are mostly dependent on grains and legumes. The utilization of available raw material such as agricultural waste reduces the dependence on the costly imported feed and banana by-products could be one of the potential raw materials for feedstock production for most of the banana producing countries.Banana peels are known to contain a substantial amount of protein, lipid, carbohydrate, fiber and a number of essential minerals such as potassium, sodium, calcium, iron and manganese, which serves as a promising raw material for feed production. Banana peels could be processed as a mycological medium for growing valuable micro-fungal biomass, enriching the protein and fatty acids content of the solid mixture. Protein content of banana peels increased up to 34 % through solid-state fermentation (SSF) by *Aspergillus niger, Aspergillus flavus* and *Pennicilium sp.* while the tremendous increase of sugar content (142 %) is possible through fermentation by *Aspergillus flavus*. Enhancement of nutritional value of banana waste through microbial fermentation is an important step in order to create high nutritional quality feedstock from low quality materials; protein and sugar content can be increased comparable to the soybean meal, which is a common ingredient in most animal feeds. Banana leaves are the best fodder because of its low partition factor, high ATP and high microbial biomass.

Banana Plant Used as An Edible Package

The natural fibres contained within them may be used in the production of rotationally moulded plastics, which are used to make everyday items such as, oil tanks, wheelie bins, water tanks, traffic cones, plastic dolls and many types of boats. The banana plant fibres will be processed, treated and added to a mix of plastic material and sandwiched between two thin layers of pure plastic providing excellent structural properties.

Liquid Fertilizer

Banana sap is a very good source of essential plant nutrients for this reason it could be used as liquid fertilizer and can be applied through fertigation (MIS) or by foliar method of application. The feasibility of using banana pseudostem sap as a liquid fertilizer for onion cultivation under drip revealed that the application of banana pseudostem sap @ 2000 l/ha through micro irrigation system increase the plant height, leaf area per plant, bulb weight, average bulb weight and dry matter yield of onion as compared to control.

Ecofriendly Products

Fabrication of banana sheath for making mates, boxes and baskets, which attract the costumers to attain more attention in the market due to natural based material. This entrepreneurs in the society waste management process will create lot of employment opportunities and promote good entrepreneurs in the society.

Conclusion

The objective of doubling the farm income could be achieved through minimizing the expenditure incurred and maximizing the profit realized by adopting the right technology. Innovation in managing such a vast amount of agricultural waste or biomass is a continuous challenge and recent trends favor the utilization of this biomass for value added purposes to fulfill the need in the areas such as renewable energy, fiber composites and textiles, food alternatives and livestock feed. Smallholder farmers in developing countries need to improve their position in food value chains in order to improve their margins and as a strategy for coping with agricultural food price volatility through innovations within the chains. Value chain mappings and gross margin analysis are necessary to keep the business viable and sustainable. Weak linkages within the banana value chains with poor integration of value chain actors and minimal involvement with regional markets and high-value domestic chains such as supermarkets make the business more volatile. Value addition in terms of agro-processing at small scale levels using rudimentary techniques limiting the final product to low value markets. It is suggested to strengthening linkages within the value chains, collective marketing, penetration into high-value chains and improved processing techniques may provide a potential avenue for enhancing banana value chains in India and around the world.

In comparison to the synthetic fibres the natural fibres are with added advantages for their stiffness, readily availability, low cost, and with renewable properties. India is the top most banana cultivator and it has about 8.3 lakh ha under banana cultivation, every year at least half the areas under banana are getting replanted. It is therefore, potentially, one billion pseudostem are available for the extraction. With the extractable pseudostem fibre yield of 400 kg fibres per hectare, 16 crores kg worth 3000-4000 crores of fibre could be extracted which is otherwise dumped as a waste in India after harvesting the bunches. Therefore, banana biomass can be used as resources for natural fibre apart from its nutritive value. Banana fibres can be used for various purposes such as in textile, paper or handicrafts industry. With the huge potential of availability of banana fibre it is pre-requisite for the development of integrated package for promoting handcraft from banana fibre and supporting MSME enterprises. Developing replicable models across regions, having diverse and convergence of ideas, design and technology development workshops, promoting farmer producer organizations (FPOs) and business friendly atmosphere are essential for promoting and utilization of banana fibre.

20

Nutraceuticals from Citrus Fruits

Dinesh Kumar and M.S. Ladaniya

ICAR-Central Citrus Research, Nagpur-440010, Maharashtra, India

Introduction

Citrus fruits are acidic, exotic fruits with juicy, bitter, fruit segments inside. Citrus fruits commonly originated in tropical and sub-tropical regions of the world. They have long been valued as part of a nutritious and tasty diet. The flavours provided by citrus are among the most preferred in the world, and it is increasingly evident that citrus not only tastes good, but is also good for people. Citrus fruits have certain therapeutic properties such as antitumor, anti-inflammatory and anticancer, due to phytovitamins and nutrients present in them.

The term "Nutraceutical" was coined by combining the terms "Nutrition" and "Pharmaceutical" in 1989 by Dr Stephen DeFelice, Chairman of the Foundation for Innovation in Medicine. “Nutraceutical” is a marketing term developed for nutritional supplement that is sold with the intent to cure or prevent disease and thus has no regulatory definition. Hence a “nutraceutical” is any substance that may be considered a food or part of a food and provides medical or health benefits, encompassing, prevention and treatment of diseases. Such products may range from isolated nutrients, dietary supplements and diets to genetically engineered "designer" foods, herbal products and processed foods such as cereals, soups and beverages. Presently over 470 nutraceutical and functional food products are available with documented health benefits.

It is well established that citrus and citrus products are a rich source of vitamins, minerals and dietary fibre (non-starch polysaccharides). These and other biologically active, non-nutrient compounds found in citrus and their products can lead to widespread nutritional benefits across the population. Antioxidants restrict the damage that reactive oxygen free radicals can cause to the cell and cellular components and prevent diseases like atherosclerosis, cancer, inflammatory joint diseases, asthma, diabetes etc. It also contains an impressive list of other essential nutrients, potassium, folate, calcium, thiamine, niacin, vitamin B_6, phosphorus, magnesium, copper, riboflavin, pantothenic acid and a variety of phytochemicals. Pectin from citrus peel is considered as one of the best pectin quality than any other fruit. Citrus peel and rag are normally used either for pectin manufacture, or they are dried and sold as cattle feed. When the citrus fruits are squeezed, about 50% of their weight is discarded as waste peel, membrane, juice vesicles and. Then obtaining by-products from these wastes may increase the economic yield of the citrus juice industries. Citrus by-products are commonly used to fortify animal fodders, but also the citrus peels are used to obtain flavonoids, aromatic components and dietary fibre concentrates.

A. Structure

Citrus fruits are covered with a rind, or peel, to protect the pulp, or edible portion, of the fruit. The rind is made up of a cuticle on the outside, thinly covering an epidermal layer (flavedo) containing numerous oil sacs, or glands, filled with an aromatic essential oil having commercial value. A white spongy portion of parenchymatous cells, known as the albedo, lies directly beneath the flavedo. This layer contains approximately 20 percent of pectinous substances, which can be recovered in the form of citrus pectin. The inner pulp or flesh of the fruit consists of segments (locules), separated by a membrane of thin epidermal tissue and containing numerous spindle-shaped juice sacs (vesicles) and seeds. The juice sacs in each segment are attached to the segment wall and in contact with the peel by means of fine threads of varying length. The central axis (core) of the fruit is composed of a white spongy tissue similar to that found in the albedo. The core and segment membranes are collectively called the "rag" of the extracted juice.

B. Nutritional Profile of Citrus Fruits

Lemon

Importance

D-limonene supports the normal functioning of immune system. Lemon acts as an antiseptic agent by preventing the growth of pathogenic bacteria. Lemon is also an important ingredient in natural cosmetics that reduces wrinkles in the face. Lemon is also useful in treating asthma, cough, biliousness, cold, heartburn, liver complaints, fever, diphtheria and rheumatism (Brower V., 1999). Lemon helps in the production of WBC and antibodies in blood which attacks microorganism and prevent the infections. Lemon acts as an antioxidant by deactivating the free radicals and prevents diseases like stroke, cardiovascular diseases and cancer. It also lowers blood pressure and increases the level of HDL.As lemon is anti carcinogenic; it lowers the rates of colon, prostate and breast cancers.

Orange

In India it is grown in places like Assam, Madhya Pradesh, Maharashtra, Meghalaya, Rajasthan, Mizoram and Karnataka.

Importance

Orange peel serves cosmetic purpose like scrubs, as it provides freshness to our skin. Natural oils from orange peels are used as excellent cleaners .It can also be used in the weight control of the body (Eskin N A M *et al.*, 2006 & Garg A *et al.*, 2001). As the orange peels are rich in vitamin A and C they fight against infections, cold and flu. In gardening, peels makes the surrounding an effective method of getting rid of slugs. It can reduce the risks of heart diseases, diabetes, asthma and osteoarthritis. These may reduce the prevention of high blood pressure or hypertension.

Grapefruit

Importance

Uplifts and energizes and gives nourishment to skin (Hatton *et al.*, 1977). It supports immune system of body. Grapefruit used to support body's natural defence. It is added to foods and drink to enhance their flavour of foods. Their fragrances reduce depression from our mind and hence When inhaled, it gives freshness to our body. Grapefruit helps in reducing weight in the body. It prevents arthritis and works as an antiseptic. Grapefruit helps in cancer prevention and lowers the cholesterol level. It also treats common treatment of cold and fever.

C. Phytonutrients

Antioxidants

Citrus phenolic compounds particularly flavonoids have been reported to possess an important antioxidant activity toward radicals. The citrus flavonoids have the ability to capture electrons, block and/or scavenge the radicals. The citrus flavonoids form a tautomeric dislocation, which prevents the propagating chain reactions of these oxygenfree radicals.

Antioxidant activity can be determined by following methods-

1. ABTS (2,2 Azinobis 3-Ethyl Benzthiozoline-6-Sufonate) method
2. DPPH (1, 1 Diphenyl 2- Picryl Hydrazyl) method
3. FRAP (Ferric Reducing Antioxidant Power) method

These three assays are used to measure the antioxidant, radical scavenging activities of juice extract.

Vitamin

The citrus fruit juice contained ascorbic acid, niacin, riboflavin and thiamin in varying quantities. Ascorbic acid in the body aids in iron absorption from the intestines. It is required for connective metabolism especially the scar tissue, bones and teeth. It prevents muscle fatigue and scurvy that is characterized by skin hemorrhages, bleeding gums, fragile bones, anemia and pains in joints and defects in skeletal calcification. It acts also as antioxidants in the skin by scavenging and quenching free radical generated by ultra violet radiation stabilization. The production of collagens is also dependent on vitamin C.

Carotenoids

Another group of phytonutrient found in citrus is carotenoids. They are referred to be the orange-red colors found in citrus. Citrus fruits contain high levels of other carotenoids (lutein, zeaxanthin, cryptoxanthin) that have significant anti-oxidant activity. Carotenoids are important in human health.

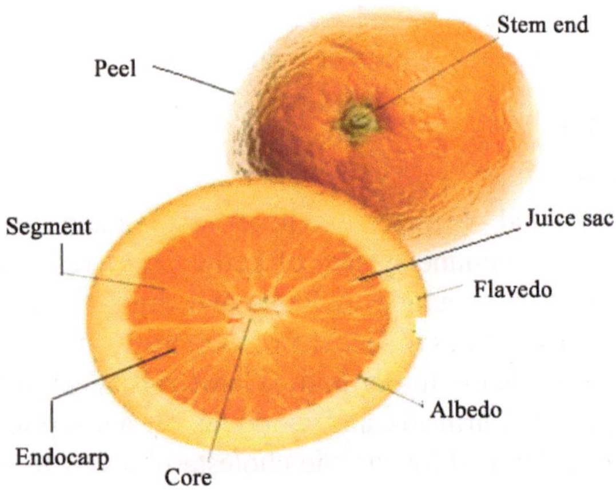

Dietary Fibre

Citrus fiber has a biochemical effect on the absorption and re-absorption of cholesterol and bile acids, respectively. Citrus soluble fiber binds to bile acids in the small intestine, making them less likely to enter the body; this in turn lowers cholesterol levels in the blood. Soluble fiber also attenuates the absorption of sugar, reduces sugar response after eating, and normalizes blood lipid levels. Insoluble fiber is associated with reduced diabetes risk.

Pectin

Pectin is a soluble fibre. It is beneficial to human health by regulating blood sugar levels and reducing serum cholesterol. Pectin is a polysaccharide widely used in food and pharmaceutical industries. It is used as thickening and gelling agents. Its medical uses are antidiarrhea, detoxification and blood glucose lowering. The carboxyl group of galacturonic acid can be free or methyl-esterified which classifies pectin into high- and low- methoxyl types depending on their degree of esterification. Nagpur mandarin peel gives a significant amount of pectin whereby it can be considered in commercial production of pectin alongside with other citrus sources.

Flavonone Glycosides

Hesperidin is a flavanone glycoside which is classified as a citrus bioflavonoid. Sweet oranges *(Citrus sinensis* L. Osbeck) and tangelos are the richest dietary sources of hesperidin. The peel and membranous parts of lemons and oranges have the highest hesperidin concentrations. Therefore, orange juice containing pulp is richer in the flavonoid than that without pulp. Hesperidin, in combination with a flavone glycoside called diosmin, is used for the treatment of venous insufficiency and hemorrhoids. Hesperidin, rutin and other flavonoids are reported to possess analgesic and anti-inflammatory activity (Nelson NJ.1999).

Although hesperidin is insoluble in water and is tasteless, naringin (in grapefruit) is soluble in hot water and has an extremely bitter taste. The aglycone of naringin, naringenin, does not have a bitter flavour. Suggestions have been made that these compounds may act as light filters to regulate photosynthesis in citrus plants.

Limonoids

Limonoids are a group of chemically related triterpene derivatives present in citrus fruit. They possess the ability to inhibit tumour. D-limonene supports the normal functioning of immune system. It gives clarity and freshness to our mind. It releases us from tension and makes our mind peaceful. It also supports and refreshes our skin.

Phenols

Polyphenols are simple phenolic molecules to highly polymerized compounds with molecular weights of greater than 30,000 Da. Plant phenol includes a variety of compounds which are structurally related. It includes the simple phenols; the hydroxyl cinnamic acid derivatives; the flavonoids including catechins (flavanols), anthocyanins, flavones and flavanols. They show effects including anti oxidant, anti-viral, anti-allergic, anti-inflammatory anti-proliferative and anti- carcinogenic. Higher phenolic contents which indicate potent antioxidant ability of pummelo was reported by Kumar *et al.* (2015).

Organic Acids

Citrus fruits are classed as acid fruits. The acidity of citrus juices is due primarily to their content of citric and malic acids. Trace amounts of tartaric; benzoic, and succinic acids as well as oxalic and formic acids have also been reported. Citric acid accounts for the largest portion of the organic acids in citrus fruits.

Conclusion

Citrus fruits are the main fruit trees grown throughout the world and are well– appreciated for their refreshing juice and health benefits. The numerous health benefits of citrus fruits are linked to the high amounts of neutraceuticals (phytochemical and bioactive compounds) such as flavonoids, carotenoids, vitamins and minerals available in citrus fruits. These neutraceuticals may act as antioxidants, stimulate the immune systems; block the damage of the genetic materials. These citrus fruits have anti-cancer, anti-septic and anti-oxidant properties. They also provide natural refreshment to the skin.

Because of growing acceptance of nutraceuticals, India could hope to leverage the country's key resources in this area to gain a foothold in the global market. Nutraceuticals are among the new age drugs that are being developed to provide better health. Nutraceuticals are gaining public acceptance in many developed countries in recent times. Looking at the changing trends, the market of nutraceuticals has huge potential. These days, industries are showing interest in the nutraceutical area. Within few years this potential can turn into a healthy growing market.

References

Brower V., 1999. Nutraceuticals: poised for a healthy slice of the healthcare market? Nat Biotechnol. 16: 728-731.

Eskin N A M and Tamir S., 2006. Dictionary of Nutraceuticals and Functional Foods, CRC Press, Boca Raton, USA.

Garg A, Garg S, Zaneveld LJD and Singla AK., 2001. Chemistry and pharmacology of the citrus bioflavonoid hesperidin. Phytother Res.; 15: 655-669.

Hatton, T.T. and R.H. Cubbedge, 1977. Status of controlled-atmosphere storage research of citrus fruit. In: D.H. Dewey (ed.) Proc. 2 Nd Natl. Contr. Atmos. Res. Conf., Mich. State Univ. Hort. Rpt. No. 28: 250-259.

Kumar D., Lamers H., Singh IP, MS Ladaniyal and Bhuwon Sthapit, 2015. Phytochemical evaluation of pummelo fruits (*Citrus grandis*) in India for enhancing marketing opportunities, Indian J. Plant Genet. Resour. 28(1): 50-54.

Nelson NJ., 1999. Purple carrots, margarine laced with wood pulp? Nutraceuticals move into the supermarket. J Natl Cancer Inst. 91: 755-757.

21

Technical and Legal Perspective of Food Processing Plant

S.K. Sharma, Deepa Saini, Neha Rawat and Riya Barthwal

Department of Food Science and Technology, College of Agriculture, GB Pant University of Agriculture and Technology, Pantnagar Uttarakhand

Food Business Regulatory System in India

In India Food Businesses are regulated by only One authority i.e. Food Safety and Standards Authority of India (FSSAI). This comes under the control of Ministry of Health and Family Welfare of the Govt. of India.

FSSAI – Food Safety and Standards Authority of India

The FSSAI or Food Safety and Standards Authority of India was incorporated under the Food Safety and Standards Act (FSSA) of 2006. It consolidated various acts & orders that were previously applicable and used to handle food-related issues in different Ministries and Departments. The basic objective of formulating FSSAI has been to lay down science-based standards for items of food and to regulate their manufacture, storage, sale, distribution, and import to ensure the availability of adulteration-free and wholesome food for human consumption.

One Nation One Food Law

To bring consistency across all the States, in enforcing food safety regulations and standards, the FSSAI initiated its "One Nation One Food Law" program. The regulator has leveraged infotech for implementation of identical standards for testing and sampling throughout various States and labs, in India

- Ensure that investors and food businesses have a consistently positive experience across all States while dealing with the regulatory environment
- Place a single reference point for all issues connected to food safety and standards, by moving from multi-departmental, multi-level control to a single line of command

- Consolidate various laws
- Regulate the manufacture, storage, packing, distribution, sale and import of food products
- Ensure the availability of contamination-free and wholesome food for human consumption

Goal of Establishing the Authority

FSSAI has been mandated by the FSS Act, 2006 to achieve the following purposes:

- Frame Regulations to set the Standards and Guidelines about food articles and specifying an appropriate system of enforcing different standards thus introduced.
- Establish mechanisms and guidelines for accreditation of certification bodies engaged in certifying efficient Food Safety Management System (FSMS) for food businesses.
- Establish procedures and guidelines for accreditation of laboratories and notification of the laboratories, accredited thus.
- Provide scientific guidance and technical assistance to different Governments and help in devising policy and rules in areas that have a direct or indirect bearing of food safety and nutrition.
- Collate and organize data about food consumption habits, scope, and prevalence of biological risks, contaminants in food, and residues of various contaminants in food products. Identification of rising risks and the induction of a rapid alert system.
- Formulate a network for information across the country so that the public, consumers, and Panchayats, may receive fast, credible and honest information about food safety and other issues of concern.
- Provide training programs to persons who are already involved or plan to get involved in food businesses.
- Contribute in the fields of development and supporting of international technical standards for food, sanitary and phyto-sanitary standards.
- Promote public awareness about food safety and standards.

Establishment of a Food Processing Plant

FSSA 2006 mentions in its Schedule 4, the requirements of FSS (Licensing & Regulation of Food Businesses) Regulation, 2011. Some of these requirements have been detailed ahead.

Selection of Location and site

Manufacturing / processing / packaging premises shall be located away from sources of pollution like open drains, garbage yards, industries that emit fumes or gases and dense vegetation.

In order to avoid contamination from external sources such as odor, pests, dust etc., appropriate measures shall be taken to protect the processing area from environmental contamination. Processing area shall not have direct access to any residential area.

Building Design and Layout

a) Outside the food production premise

Layout & Design of the food production unit should be uni-directional to prevent backward flow of materials during processing. This is required in order to prevent cross contamination.

b) Inside the food production premise

i) Floors, ceilings and walls of the establishment must be made of impervious material. They should be smooth and easy to clean with no flaking paint or plaster and maintained in a sound condition to minimize accumulation of dirt, condensation & growth of moulds.

ii) The doors in the establishment shall be made of smooth and non-absorbent surfaces and they shall be easy to clean and disinfectant. Doors can be fitted with automatic closing spring, strip or air curtain.

iii) The floor shall have adequate and proper drainage with appropriate slope and they should be easy to clean and disinfect. The drainage shall flow in a direction opposite to the direction of food preparation area to avoid contamination

iv) The drains should be covered to prevent insects and rodents from entering the processing area

v) The windows, doors & all other openings to outside environment in the establishment shall be well screened with wire-mesh or insect- proof screen to protect the premise from pests. The doors shall be fitted with automatic closing springs to keep them closed at all times and also the mesh should be easy to remove & clean to avoid accumulation of dust & dirt.

Design of Equipment and Plants

During preparation / handling / processing / storage of food products like flour, sugar, dairy based products, spices, bakery ingredients, dry fruits, cereals

& grains, oilseed, refined oil, etc. certain equipment's e.g. scoops, spoons, cooking vessels, containers, tanks, silos, hoppers, pipes, packaging machines, filters, etc. come into contact with food. All these food contact surfaces shall be:

- Made up of non-corrosive / rust free material
- Smooth, free from any grooves
- Easy to clean and maintain
- Non-toxic and non-reactive
- Of food grade quality

Generally, food grade stainless steel or galvanized iron material is preferred as it complies with the above requirements.

- All the equipment's used during food preparation, handling &packaging (if any) shall be designed, located and fabricated to facilitate easy cleaning and shall be kept away from impure air anddust.
- Every utensil or container containing food products shall at all times be either provided with a properly fitted cover/lid or with a clean gauze net or other material of texture sufficiently fine to protect the food completely from dust, dirt and flies and other insects.
- No utensil or container used for the manufacturing of food items shall be kept in any place in which such utensil or container be contaminated and thereby render the food noxious.
- All equipment shall be kept clean, repaired and maintained in sound condition all the time.

All measuring instruments / equipment like temperature gauges, pressure gauges, weighing balances, etc. shall be calibrated periodically for correct measurement

FSSAI License – Requirements and Procedures

For undertaking any kind of food business in India, the Food Business Operator has to obtain FSSAI license. The application can be submitted through FLRS (Food Licensing and Registration System) the FSSAI portal for this purpose. FBOs have to move to FoSCoS (Food Safety Compliance System) w.e.f. 1st Nov 2020, which shall replace the existing FLRS. The details of the types of licenses, fee, documents requirements have been discussed in this section.

Types of Licenses

Food Business Operators (FBOs) involved anywhere in the food supply chain need to apply for food licenses of different types to different authorities. This categorization is based on their turnover, scale of business and the kind of activity. Such businesses/units must apply for either of the Licenses (Central or State) or simply the Basic Registration. The rules are as follows:

Registration

This kind of Food Registration is to be applied to the State in which the business unit is located. Small FBOs or most Start-Ups are covered under this. Criteria for eligibility are as below for those units whose turnover does not exceed □ 12 lakhs per annum:

- The production capacity of food (excluding milk or milk products and meat or meat products) does not exceed 100 kg/litres per day, or,
- Procurement or handling & procurement of milk is up to 500 litres of milk per day, or,
- Slaughtering capacity of 2 large animals or 10 small animals (including sheep and goats) or 50 or lesser poultry birds per day.

FSSAI State License

This is, generally, required by the mid-sized companies and units. Criteria are as follows:

- Turnover anywhere between Rs. 12 lakhs to 20 crores per annum, or,
- Production capacity of food (other than milk or milk products and meat or meat products) of up to 500 kg/litres per day, or,
- Production capacity of food (other than milk or milk products and meat or meat products) of up to 500 kg/litres per day, or,
- Procurement or handling & procurement of milk is up to 5000 litres of milk per day, or,
- Slaughtering capacity of anywhere between 2 to 50 large animals or 10 to150 small animals or 50 to up to1000 poultry birds per day

FSSAI Central License

This is required by large businesses or units. Criteria for eligibility are:

- Turnover of above Rs 20 Crores per annum, or,
- Dairy units (including milk chilling units) equipped to handle or treat over 50,000 litres of liquid milk daily or 2500 metric tonnes of milk solid annually, or,
- Vegetable oil-producing units, processing units, and refineries (including oil expeller units) with an installed capacity of over 2 metric tonnes (MT) daily, or,

- All slaughterhouses equipped to slaughter more than 50 large animals or 150 or over small animals or 1000 or over poultry birds daily, or,
- Meat processing units equipped to handle or treat more than 500 kg of meat daily or 150 MT annually,
- 100 % Export Oriented Units (EOUs), or,
- Importers who wish to import food items (incl. food ingredients and additives) for commercial use, or,
- FBOs operating in more than one State.

Fee Structures

For obtaining FSSAI license, FBOs have to deposit annual license fee as detailed ahead for different categories. In case of registration and state licenses, one time payment of license fee for a period of 1-5 years is also allowed.

Table 1.1 Details of applicable license fee for registration and central licensing

Purpose	**Central License (1 Year)**	**State License (1 Year)**	**Registration (1 Year)**	**Konkan / Indian Railway (1 Year)**
New Application	□ 7500	Variable	□ 100	□ 2000
Renewal Application	□ 7500	Variable	□ 100	□ 2000
License / Certificate Modification	□ 7500	Variable	□ 100	□ 2000
Duplicate License / Certificate	10 % of the applicable license fee	Variable	10 % of the applicable license fee	10 % of the applicable license fee

Table 1.2 Details of applicable license fee for state licensing

Manufacturer / Miller	**State License Fee (1 Year)**
Above 1 MT per day production	□ 5000
10,0001 to 50,000 LPD of milk or 501 to 2500 MT of milk solids per annum	
Below 1 MT per day production	□ 3000
501 to 10,000 LPD of milk or 2.5 to 500 MT of milk solids per annum	
Hotels upto 4 star	□ 5000
All food service providers including restaurants / boarding houses, clubs etc. serving food, canteens (schools, colleges, office, institutions), Caterers, Banquet hall with food catering arrangements, food vendors like dabba walas etc& other food business operators	□ 2000
Renewal of License / Certificate	Depends on No. of years

Modification of License / Certificate	Corresponding fee for 1 year
Duplicate License / Certificate	10 % of the applicable license fee

Procedures and Documents Required

This section contains the details of application procedures for different types of licenses and the requisite documents for each of these categories.

Basic Registration

Procedure

- Every small FBO shall register itself with the Registering Authority (RA) by submission of an application form, along with the fees as applicable.
- A Unique Application Reference Number would be generated.
- RA may grant or reject the application – with reasons being recorded in writing. Within 7 days of the receipt of the application.
- On the grant, the FBO may commence the business operations.
- Or, if decided necessary, the RA may issue a report for inspection. This is ordered within 7-days. And the inspection is conducted within 30 days.
- If through the inspection, the inspecting FSO or agency is satisfied, the FSSAI number will be granted.

Documents Required for Registration

Listed below are the common documents mandatorily required to be submitted to get an FSSAI any type of Food License or Registration. The documents that are necessary to be submitted by all FBOs, have been mentioned below, whichever type of Certificate you are applying for Basic Registration, State or Central License.

i. Form A or Form B (Form – A for Basic FSSAI Registration & Form – B for State and Central FSSAI Registration) – completed and signed,

ii. 2 Passport Sized Latest Photographs,

iii. PAN Card of the Business Owner/Proprietor/Partner/Director,

iv. Proof of Possession of the Premises (Property Papers – for owned property. Utility bills (Electricity or water) or Rent Agreement – if the premises is on rent),

v. Partnership Deed,

vi. Certificate of Incorporation or Articles of Association (AOA) of a company,
vii. The complete list of food items that the unit will be processing,
viii. The planned FSMS (Food Safety Management System). Applicable for FSSAI State or Central Food License only.
ix. Authority letter with name and address of responsible person nominated by the manufacturer and an alternative responsible person indicating the powers vested with them. Such as, assisting the officers in inspections, collection of samples, packing & dispatch. This is mandatory for manufacturers and processors, (applicable for Licenses only).
x. Analysis report (Chemical & Bacteriological) of water to be used as an ingredient in food from a recognized/public health laboratory to confirm the portability. Mandatory for manufacturing and processing units. (applicable for Licenses only).

State license

Procedure

i. Filing of Application Form no. B. All associated documents attached and fees paid, as applicable
ii. Unique Application Reference Number is generated on this
iii. Now, there are 3 ways possible:
   a. If FSSAI is content with the details provided in the application and the documents submitted, the FBO can commence its business operations
   b. Or additional documents may need to be submitted or some other information, missing in the application, may be required. On submission, a unique application ID gets generated on this. Now, if no inspection is called for, the FBO may begin operations after 60 days. However, if an inspection is required, it would be conducted and a report issued. Now, the license will be given if all the formalities have been complied with, within 60 days. If in case, the inspection report is not processed, the FBO may begin the business after 60 days
   c. Or the FBO gets an Improvement Notice. It will also mention a period within which the case is to be modified. If the issue is not resolved within that time period, the application would be held under suspension. And another date will be announced along with a penalty payable. If the issue is not revised still, the application will get cancelled. A new application can be filed only after 90 days from this cancellation.

Documents for State Food License

Below documents need to be attached with the application, other than the above-listed essential ones:

i. A completed and signed Form no. B,
ii. Details of the Business Owner/Proprietor/Partners/Directors. And their Photo ID, Address Proofs, and Photographs,
iii. The planned layout of the location, and the proposed placement of the equipment,
iv. Possession Proof of the unit/premises,
v. No-Objection-Certificate (NOC) from the premises owners. And their details,
vi. A copy of License from the manufacturers of the Equipment,
vii. List & Details about the Equipment already installed or planned to be installed,
viii. Food Products and their categories that are proposed to be handled.

Central License

Procedure

i. Filing of Application Form no. B. All associated documents attached and fees paid, as applicable
ii. Unique Application Reference Number is generated on this
iii. Now, there are 3 ways possible:
   a. If FSSAI is content with the details provided in the application and the documents submitted, the FBO can commence its business operations
   b. Or additional documents may need to be submitted or some other information, missing in the application, may be required. On submission, a unique application ID gets generated on this. Now, if no inspection is called for, the FBO may begin operations after 60 days. However, if an inspection is required, it would be conducted and a report issued. Now, the license will be given if all the formalities have been complied with, within 60 days. If in case, the inspection report is not processed, the FBO may begin the business after 60 days
   c. Or the FBO gets an Improvement Notice. It will also mention a period within which the case is to be modified. If the issue is not resolved within that time period, the application would be held under suspension. And another date will be announced along with a

penalty payable. If the issue is not revised still, the application will get cancelled. A new application can be filed only after 90 days from this cancellation

Documents for Central Food License

i. Other than the above-listed Common Documents, you need below:
ii. A completed and signed Form no. B,
iii. Details of the Business Owner/Proprietor/Partners/Directors. And their Photo ID, Photographs and Address Proofs,
iv. Proof of Annual Turnover,
v. Layout of the proposed premises/unit. And the planned placement & installation of the equipment,
vi. Proof of Possession for the premises/unit,
vii. No-Objection-Certificate (NOC) from the premise's owners. And their details, such as name & address,
viii. Copy of License from the manufacturers of the Machinery & Equipment,
ix. List and Details of the equipment planned to be installed or already installed at the premises,
x. All Food Products with their categories that are proposed to be handled,
xi. Proposed System for Food Safety Management (FSMS),
xii. Water Report Analysis from a Government Accredited Laboratory,
xiii. Source of procuring Milk, Meat & its processing unit, if applicable,
xiv. NOC (No-Objection-Certificate) from the Local Authority, if applicable,
xv. If the operations are about Mineral or Carbonated water, then Pesticide Residual Report by a Government Accredited Laboratory,
xvi. Import and Export Code (IEC), as applicable,
xvii. Certificate from the Tourism Ministry, if applicable,
xviii. Proof of Turnover for Vehicles, if applicable,
xix. No-Objection-Certificate (NOC) from FSSAI.

Non-Compliance

All businesses engaged in activities related to food, have to comply with the rules, as introduced and updated from time to time, by the FSSAI. These are statutory obligations that an FBO must meet. Food Authority conducts random inspections and tests the food articles or the process. During the inspection, the Food Safety Official (FSO) or the agency is authorized to examine the business premises, safety, and quality of the process or the food item. Or even the level of hygiene maintenance. He can check the working conditions along with the process of handling food by the operators, associates, and organization.

A level of compliance for the FBO is ascertained, using a checklist, based upon the provisions of the FSSAI. Using the checklist, the FSO marks a level for the FBO, as either:

- Compliance (C),
- Partial compliance (PC),
- Non-compliance (NC), or,
- Not applicable/Not observed (NA)

Thereby, a Notice for Improvement may be issued by the FSO or the inspection agency. It explains the place or steps where modifications need to take place. And a time duration is stipulated for these changes to be finished by. If the FBO fails to incorporate these adjustments, within the appointed time, a Show Cause Notice is given.

This Show Cause Notice provides a date of hearing. On which, the FBO has to explain why it could not make the improvements or changes. Documentary proof must be provided with the claim. If the proof gets rejected, or the FBO fails to reply to the Notice, by the hearing date, its FSSAI application will get cancelled.

If the FBO has reasons to believe that the Notice of Improvement is biased or untrue, it can Appeal to the State Commissioner for Food Safety.

The decision of the State Commissioner is also not final. FBO can dispute it and approach the Food Safety Appellate Tribunal or the High Court to file the subsequent appeal. The decision of the tribunal or the court, however, is considered final.

Penalties

If an FBO is observed Non-Compliant with the requirements of the FSSAI, it will get penalized. The penalty structure is constant and applies to everyone. Some of the FSSAI penalties for a non-compliant FBO are listed below:

- If an entity is operating without applying for the Food Safety certificate, this is punishable under the Act. The FBO may face custody and/or a penalty of up to Rs. 5 lakhs.
- If someone is selling sub-standard quality food product, a fine of up to Rs. 5 lakhs may be imposed. And the person responsible for the sales, either himself or by a person operating on his behalf, would be held liable.

- The FBO selling mis-branded food items, either himself or by a person acting on his behalf, may be imposed a fine of up to Rs. 3 lakh.
- For all the businesses selling/packing/storage/distributions or importing of Food Items containing superfluous or extraneous substance, a fine of Rs. 1 lakh may be imposed.

The fines and penalties listed above may vary on a case to case basis.In extreme cases, where death has occurred due to consuming that particular Food Item, the punishment to the responsible FBO may extend to Rs. 5 lakh or even imprisonment, depending upon the gravity of the situation and how much harm has been caused.

Food Safety Management Systems

The food safety management system is a tool to ensure your food safety at food production premises. The Food Safety and Management System (FSMS) is a set of standards established to direct and control food safety aspects. Also referred to as FSMS, it helps the food business operators to gain this trust of the consumers or even serve them efficiently. A food business organization that beholds certification for FSMS depicts assurance that the organization has taken care of appropriate Food Safety and Management System.

Steps for Implementation of Food Safety Programme

i. Implement 8 principles of PRPs
ii. Implement 6 preliminary steps of HACCP
iii. Implement 7 principles of HACCP
iv. Review HACCP plans

Pre-requisite Program (PRP)

Defined by ISO as the basic conditions and activities necessary to maintain a hygienic environment throughout the food chain which are suitable for the production, handling and provision of safe end products and safe food for human consumption.

Pre requisite progs are those which provide basic environment and operating conditions for production of safe food. These are a prerequisite to a successful implementation of HACCP. eg GHP, GMP.

There are a wide variety of PRPs depending on the particular product and process. They are often described in certain sectors of the industry as Good Agricultural Practice (GAP), Good Veterinarian Practice (GVP), Good Manufacturing Practice (GMP), Good Hygienic Practice (GHP), Good

Production Practice (GPP), Good Distribution Practice (GDP), and Good Trading Practice (GTP).

Good Manufacturing Practices (GMP)

GMPs or current Good Manufacturing Practrices (**cGMPs**) are the industry-specific 'good manufacturing practices' for food products. These practices are the fundamental operational requirements necessary to enable a food business to produce safe food. They are about the methods, equipment, facilities, and controls for producing safe and wholesome food. They are an important part in HACCP, ISO22000, and any other food safety system. Examples of GMP elements:

- Pest control
- Water supply
- Rubbish and internal organ disposal
- Personnel hygiene
- Warehousing and distribution

Good automated manufacturing practice (**GAMP**) – is a set of guide-lines for manufacturers and users of automated systems in the pharmaceutical industry.It is a technical subcommittee of the International Society for Pharmaceutical Engineering (ISPE).

Good Hygiene Practices (GHP) – system/ measures for maintaining minimum hygiene and sanitation to ensure that food is safe and suitable for human consumption. It includes personal hygiene, employee health conditions, maintenance of plant & equipment hygiene including food contact surfaces, pest control, waste disposal, water quality, toilet & hand wash facilities, prevention of cross contamination etc.

Good agricultural practice (**GAP**) – is a certification system for agri-culture, specifying procedures and documentation that must be implemented to create food for consumers or further processing that is safe and wholesome, using sustainable methods. While there are numerous competing definitions of what methods constitute good agricultural practice there are several broadly accepted schemes that producers can adhere to.

GAPs may be applied to a wide range of farming systems and at different scales. They are applied through sustainable agricultural methods, including economically and efficiently produce being sufficient (food security), safe (food safety), and making sure that the food is nutritious (food quality).

Examples of Pre-Requisite Programmes (PRP's) in a food manufacturing environment would be:

- Construction and layout of buildings and associated utilities
- Lay-out of premises, including workspace and employee facilities
- Supplies of air, water, energy and other utilities
- Supporting services, including waste and sewage disposal
- The suitability of equipment and its accessibility for cleaning, maintenance and preventative maintenance
- Management of purchased materials (e.g., raw materials, ingredients, chemicals and packaging), supplies (e.g., water, air, steam and ice), disposals (e.g., waste and sewage) and handling of products (e.g., storage and transportation)
- Measures for the prevention of cross-contamination
- Cleaning and sanitizing
- Pest control
- Personnel hygiene

PRP's are usually general to the process and not focused on any particular step in the process. For example, cleaning and sanitizing are activities which can apply to all steps, rooms, items and building fabric. They also have the character that their failure does not necessarily lead to an immediate and imminent food safety risk.

Failure of a PRP does not necessarily lead to an immediate and imminent food safety risk. Generally, time and repeated failure are required to create a critical change in the safety of the product.Stated another way, they usually manage more general and lower risk hazards.

Operational Pre-Requisite Programmes (OPRP)

As a concept of control, it was introduced by the ISO in their food safety management standard ISO 22000 and is defined as a control measure identified by the hazard analysis as essential in order to control the likelihood of introducing food safety hazards and/or the contamination or proliferation of food safety hazards in the product(s) or in the processing environment.

What is striking about this control measure and its definition is the almost universal confusion users have in differentiating it from CCP's and PRP's. The variation in the focus of the control measures appears to be based on a subtle difference in the description of hazards and risks.An OPRP is, specific action relating to the process, and, while not being critical for food safety, it is essential in reducing the likelihood of a specific hazard occurring.

SSOP

A Sanitation Standard Operating Procedure (SSOP) is a written document of procedures or programs used to maintain equipment and the environment in a sanitary condition for food processing. It is a step-by-step description of cleaning and sanitizing procedures and specifies

- What is to be cleaned
- How it is to be cleaned,
- How often it is to be cleaned, and
- What records are used to monitor the procedures.

SSOP is a fundamental part of a Food Safety Plan. It may be a stand-along procedure or may be a Pre-Requisite Programmes (PRP's). It shall be updated whenever there is a change in processes or chemicals used. It should be reviewed annually with the Food Safety Plan. An SSOP may written for

- A piece of equipment,
- Several pieces of equipment in a process,
- An environmental area,
- As a Master Sanitation Plan for the whole facility.

HACCP (Hazard Analysis and Critical Control Points)

HACCP is a systematic preventive approach to food safety and pharmaceutical safety that identifies physical, allergenic, chemical, and biological hazards in production processes that can cause the finished product to be unsafe, and designs measurements to reduce these risks to a safe level.

Key points

- HACCP is a FSMS
- It provides structure for objective assessment of 'what can go wrong' and requires controls to be put in place to prevent problems.
- HACCP focuses on *prevention of hazards* rather than finished product inspection
- The HACCP system can be used at all stages of a food chain, from food production and preparation processes including packaging, distribution, etc.
- The use of HACCP is currently voluntary in other food industries.
- It originated as part of the USA's manned space programme
- It is recognised internationally as the most effective way to produce safe food

- The HACCP principles apply a logical and common sense approach to food control
- Because it is a step by step approach it is less likely that hazards will be missed. HACCP, therefore, offers increased confidence to the food business and its customers.
- HACCP is cost effective through prevention of waste and incident costs.
- HACCP helps to demonstrate due diligence where required.

Key words

i. **Control point** - Any point, step or procedure at which biological, physical, or chemical factors can be control

ii. **Corrective action** – Procedure to be followed when a deviation occurs

iii. **Critical control point**-A point, step or procedure at which control can be applied and a food safety hazard can be prevented, eliminated or reduced to acceptable levels

iv. **A Critical limit**- A criterion that must be met for each preventive major associated with a critical control point

v. **Deviation** – Failure to met a critical limit

vi. **Hazard** – A biological, chemical, physical property that may cause a food to be unsafe for consumption

vii. **Risk** – An estimate of the likely occurrence of the hazard

viii. **Food borne illness**- Disease carried to people by food

ix. **Potentially hazardous food** – Natural or synthetic food supporting the rapid and progressive growth of micro organism

Priliminary Steps of HACCP

1. Define terms of reference (clearly spells out scope, extent, resources committed, terms of appointment of outside consultants, third party auditors, levels of authority and accountability for implementation).
2. Select the HACCP team (include product and process team members).
3. Describe the product and process
4. Identify intended use
5. Construct a flow diagram (schematic flow that describes the process)
6. On-site confirmation of flow diagram (walk through).

The HACCP seven principles

Principle 1: Conduct a hazard analysis. Plans determine the food safety hazards and identify the preventive measures the plan can apply to control these hazards. A food safety hazard is any biological, chemical, or physical property that may cause a food to be unsafe for human consumption.

Principle 2: Identify critical control points. A critical control point (CCP) is a point, step, or procedure in a food manufacturing process at which control can be applied and, as a result, a food safety hazard can be prevented, eliminated, or reduced to an acceptable level.

Principle 3: Establish critical limits for each critical control point. A critical limit is the maximum or minimum value to which a physical, biological, or chemical hazard must be controlled at a critical control point to prevent, eliminate, or reduce to an acceptable level.

Principle 4: Establish critical control point monitoring requirements. Monitoring activities are necessary to ensure that the process is under control at each critical control point. In the United States, the FSIS is requiring that each monitoring procedure and its frequency be listed in the HACCP plan.

Principle 5: Establish corrective actions. These are actions to be taken when monitoring indicates a deviation from an established critical limit. The final rule requires a plant's HACCP plan to identify the corrective actions to be taken if a critical limit is not met. Corrective actions are intended to ensure that no product injurious to health or otherwise adulterated as a result of the deviation enters commerce.

Principle 6: Establish procedures for ensuring the HACCP system is working as intended. Validation ensures that the plants do what they were designed to do; that is, they are successful in ensuring the production of a safe product. Plants will be required to validate their own HACCP plans. FSIS will not approve HACCP plans in advance, but will review them for conformance with the final rule.

Verification ensures the HACCP plan is adequate, that is, working as intended. Verification procedures may include such activities as review of HACCP plans, CCP records, critical limits and microbial sampling and analysis. FSIS is requiring that the HACCP plan include verification tasks to be performed by plant personnel. Verification tasks would also be performed by FSIS inspectors. Both FSIS and industry will undertake microbial testing as one of several verification activities. Verification also includes 'validation' – the process of finding evidence for the accuracy of the HACCP system (e.g. scientific evidence for critical limitations).

Principle 7: Establish record keeping procedures. The HACCP regulation requires that all plants maintain certain documents, including its hazard analysis and written HACCP plan, and records documenting the monitoring of critical control points, critical limits, verification activities, and the handling of processing deviations.

12 Steps of HACCP

1. Assemble HACCP Team
2. Describe Product
3. Identify intended Use
4. Construct Flow Diagram
5. Onsite Confirmation of Flow Diagram
6. List all potential hazards associated with each step, conduct a hazard analysis & consider any measures to control identified hazards
7. Determine the CCPs
8. Establish Critical Limits for each CCP
9. Establish a monitoring system for each CCP
10. Establish Corrective Actions
11. Establish Verification Procedures
12. Establish documentation & Record Keeping

HACCP is internationally recognized as a logical tool for adapting traditional inspection methods to a modern, science-based, food safety system. Based on risk-assessment, HACCP plans allow both industry and government to allocate their resources efficiently in establishing and auditing safe food production practices. In 1994, the organization of *International HACCP Alliance* was established initially for the US meat and poultry industries to assist them with implementing HACCP and now its membership has been spread over other professional/industrial areas.

Hence, HACCP has been increasingly applied to industries other than food, such as cosmetics and pharmaceuticals. This method, which in effect seeks to plan out unsafe practices based on science, differs from traditional "produce and sort" quality control methods that do nothing to prevent hazards from occurring and must identify them at the end of the process. HACCP is focused only on the health safety issues of a product and not the quality of the product, yet HACCP principles are the basis of most food quality and safety assurance systems, and the United States, HACCP compliance is regulated by 21 CFR part 120 and 123. Similarly, FAO/WHO published a guideline for all governments to handle the issue in small and less developed food businesses.

International Standards and Certifications

Safety is one of the key factors in the food industry. Both consumers and health authorities demand that the food we consume offer high levels of quality and safety throughout the entire production chain and until its consumption by people. If there is a problem, it is necessary to identify its causes quickly and solve it as soon as possible. Therefore, there is a wide range of regulations and a large number of certifications that can guarantee food safety so that its consumption is suitable for certain consumer groups.

Certifications are documents that certify the fulfillment by the applicant organization of a series of requirements that may be of a **legal or regulatory nature, as well as a voluntary nature.**

ISO 9001

ISO 9001 is an internationally recognized quality management system (QMS) standard. It is a world leader in SGC, surpassing one million certificates worldwide.

This standard is applicable to any organization, regardless of its size, sector or activity, or geographic location. By focusing on processes and customer satisfaction rather than procedures, it is equally applicable to both service providers and manufacturers.

Thanks to this norm, a series of principles are transmitted as:

- Commitment to its shareholders
- Good reputation in terms of the organization
- Customer-oriented management
- Competitive advantage

ISO 22000

ISO 22000 is the international standard for food safety management systems for the entire supply chain, from farmers and ranchers to processors. It involves interactive communication, system management and prerequisite programs (PPR).

ISO 22000 focuses on securing the supply chain, has principles of integrated management systems and is aligned with the HACCP principles of the Codex Alimentarius and other ISO standards of management systems.

The most outstanding benefits when getting this certificate are:

- Facilitate compliance with legislation:
    - Regulation 852/2004 “Hygiene of food products” (HACCP)

    - Regulation 178/2002 "Traceability and Food Safety"
    - Regulation 183/2005 "Hygiene of feed" (HACCP)
- Improves consumer confidence in products.
- Incorporates greater control of risks for food safety.

FSSC 22000

FSSC 22000 is a food safety scheme based on the ISO 22000, ISO 22002 and BSI PAS 220 standards, which specify the guidelines to follow in a management system to ensure the safety of food.

The Foundation for Food Safety Certification has developed this certificate based on the UNE-EN ISO 22000 standard and the British PAS 220 specification for the certification of food manufacturers.

It is supported by the European Confederation of Food and Beverage Industries (CIA), and is approved by the Global Food Safety Initiative, GFSI (Global Initiative for Food Safety)

BRC Global Standard for Food Safety

BRC Global Standard for Food Safety is a certification standard that includes the requirements of an HACCP (Hazard Analysis and Critical Control) system in accordance to the Codex Alimentarius. This standard also covers a documented quality management system, as well as the control of requirements of the environmental conditions of the facilities, products, processes, and personnel.

BRC enjoys international recognition, and was developed with the aim of helping distributors comply with the legal obligations of food safety and guarantee the highest level of protection for the consumer.

IFS (International Featured Standards)

IFS Food was created with the aim of obtaining a common **quality and food safety standard.** In this way, food manufacturers or food packaging companies are audited to optimize resources and guarantee transparency throughout the food preparation process. Both IFS and the BRC certification are essential for food suppliers in the food distribution sector.

This standard allows access to the food markets of the United Kingdom, Germany, France and Italy.

GLOBAL GAP

GLOBAL GAP is a private organization responsible for establishing a series of voluntary standards with which you can certify products from the primary sector. These standards cover the entire production process of the product.

Some of the benefits of obtaining this certificate is that it shows customers (distributors, intermediaries, importers) that the products are made based on good agricultural practices, which also inspires confidence in consumers and guarantees access to other markets.

Halal

Halal certification is a process of quality assurance applied to foods, products and services that follow the Halal Regulations.

This certification, which is granted by the Halal Institute, guarantees that the products do not include or contain in their composition anything that is considered illegal by **Islamic law**, and that these are prepared, processed, transported and stored using means that are exempt from any element prohibited by said law. In addition, it guarantees that the products have not been in direct contact with other foods that do not meet these requirements.

KOSHER

Kosher certification proves that the products obtained respect the precepts of the **Jewish religion**, and therefore are considered pure and fit to be ingested by practitioners of that religion. It covers everything from the composition and ingredients of the product to the production process, paying special attention to the preparation and cleaning of the machinery used in the preparation.

Kosher certification is a tool for differentiation and competitive positioning at an international level, in a market characterized by constant growth. This certification is important to promote the export of food products to countries where an important Jewish community is living or living.

NO GMO

This certificate guarantees that the ingredients of the product to be analyzed are not genetically modified, that is, they are not GMO (Genetically Modified Foods)

The norm around this certification ranges from the seed, through to the process of growth and harvest, transport, collection, storage and processing in the market channel. The companies in charge of issuing the certification can certify independently of the quality management systems. The certification also includes verification of the legal requirements related to the labeling and monitoring of GMOs or in Spanish OGM (Genetically Modified Organisms)

Source: BTSA antioxidants

Certification bodies Accredited by National Accreditation Board

NABCB

National Accreditation Board for Certification Bodies

Name of the Certification Bodies	Valid From	Valid Upto
Bureau Veritas (India) Pvt. Ltd.*	13 Feb 2007	12 Apr 2022
TQ Cert Services Private Limited (formerly FOOD Cert India Private Limited) (operating food services through their division TQ Foodcert) *	17 Aug 2007	16 Jan 2023
TUV SUD South Asia Private Limited*	16 Dec 2008	15 Apr 2024
TUV India Pvt. Ltd.*	24 Mar 2009	22 Sep 2020
IRCLASS Systems and Solutions Private Limited (Operating certification services through its Division Indian Register Quality Systems)*	25 Jun 2009	24 Jun 2020
INDOCERT *	18 Jan 2010	17 Jan 2021
SGS India Pvt. Ltd. *	01 Apr 2010	31 Oct 2021
DNV GL Business Assurance India Pvt. Ltd.	21 Dec 2010	20 Dec 2021
OneCert International Pvt. Ltd.*	20 Jun 2011	19 Sep 2022
Intertek India Pvt. Ltd.*	09 Dec 2011	08 Dec 2022
Vexil Business Process Services Pvt. Ltd.	22 Jun 2012	21 Mar 2020
BSI Group India Pvt Ltd.*	18 Oct 2013	15 Dec 2020
MS Certification Services Pvt. Ltd.*	28 Mar 2014	27 Jun 2021
COTECNA Inspection India Pvt. Ltd.*	28 Mar 2018	27 Mar 2021
Standards Organisation of Nigeria	31 Dec 2018	30 Dec 2021
Integrated Quality Certification Pvt. Ltd.	29 Mar 2019	28 Mar 2022
Prime Certification & Inspection LLC	29 Mar 2019	28 Mar 2022
BSCIC Certifications Pvt. Ltd.	20 Feb 2020	19 Feb 2023

* NABCB accredited Certification Bodies allowed to issue certificates for both ISO 22000:2018 and ISO 22000:2005 standards.

Note: Certification Bodies not marked with *are allowed to issue certificates as per ISO 22000:2005 only.

Tea Board India Under Ministry of Commerce and Industry Government of India

The following five Certification Bodies have been recommended for Codex HACCP based Food Safety Management System Standards Certification (ISO 22000, RVA Dutch HACCP, BRC, FS 22000, Synerge 22000 and other accredited certificates) in tea for the year 2018-19.

1	Intertek India Pvt. Ltd. Intertek House, E-20, Block 1 Mohon Cooperative Industrial area Mathura Road, New Delhi-110 044	31.10.2020
2	DNV-GL Business Assurance India Pvt. Ltd. P 540 Raja Basanta Road Kolkata-700 029	31.10.2020
3	M/S Certification Services Pvt. Ltd. HQ: 3/23 R.K. Chatterjee Road Kolkata-700 042	31.10.2020
4	Indian Register Quality Systems IR Class System and Solution Pvt. Ltd. 205, Centre Point, 21, HemantaBasu Sarani, Kolkata- 700 001	31.10.2020
5	Cotecna Inspection India Pvt. Ltd. Neelamber Building, 1st Floor 28 B, Shakespeare Sarani Kolkata- 700 017	31.10.2020

FSSAI Initiative - Tasty Food Safe Food

List of Approved Certification Bodies

Name	**Address**	Contacts
Bureau Veritas Certification (India) Pvt. Ltd.	Marwah Centre, 6th Floor, Opposite Ansa Industrial Estate, KishanlalMarwah Marg, Off Sakivihar Road, Andheri East, Mumbai - 400 072	+91 22 6695 6316 ramesh.koregave@ in.bureauveritas.com, rahul.nayak@ in.bureauveritas.com
TQ Cert Services Private Limited	" (Formerly FoodCert India Private Limited)" A Wholly owned Subsidiary of TATA PROJECTS LIMITED. SBU Quality Services Splendid Tower, 6th Floor, H No. 1-8-364, 437, 438 and 445 S P Road, Begumpet, Hyderabad -500016, Telangana,India	+91-40-2330 1618 / 2330 1554 sampatsuri@tataprojects. com
TUV SUD South Asia Private Limited	TÜV SÜD House, Off Saki Vihar Road, Saki Naka, Andheri (East), Mumbai – 400072, India	+91 20 6754 5555 Vibhavari.Bhagwat@tuv-sud.in
SGS India Pvt. Ltd.	SGS India Private limited, Ecospace, Block - 3A, 2nd Floor, East Wing I Premises IIF/ 11, Action Area -II Rajarhat, New Town I Kolkata -700 156.	033 66266118 soumik.mondal@sgs.com

IRCLASS Systems and Solutions Private Limited (IRCLASS)	2nd Floor, New Building, 52 A, Adi Shankaracharya Marg, Opposite Powai Lake, Powai, Mumbai - 400 072.	+91 22 3051 9836 / 3051 9800 irqs@irclass.org
MS Certification Services Pvt. Ltd.	3/23, R.K. Chatterjee Road, Kolkata- 700 042	+ 91 033-2441-7657 mscertinsp@yahoo. co.in;hq@mscertification. org
BSI Group India Pvt Ltd.	The Mira Corporate Suites (A-2), Plot 1 & 2, Ishwar Nagar, Mathura Road, New Delhi - 110 065	+91 11 2692 9000 sandeep.dua@bsigroup. com;Mini.Sharma@ bsigroup.com

Accreditation Bodies

NABL : National Accreditation Board for Testing & Calibration Laboratories

NABCB : National Accreditation Board for Certification Bodies

NABH : National Accreditation Board for Hospitals & Healthcare Providers

NABET : National Accreditation Board for Education and Training

QCI : Quality Council of India

FSSAI : Food Safety and Standards Authority of India

General Procedures for Food Safety Certification

1. **Application:** The company submits application through the prescribed mode on the prescribed format along with all documents, fee etc. as required by the certifying agency. Company must appoint an authorized management representative or food safety team leader to provide all kinds of information.
2. **Application review**: Once complete application of client is received, the contract reviewer of the certifying agency will review the same on the basis of applicable standards. Once required information is gathered or requirements raised (if any) are completed, an onsite audit is planned on a mutually agreed date and time. Contract reviewer will appoint the auditor/ audit team as per the requirement of the client's management system in consultation with the Quality Assurance Executive.
3. **Stage 1 audit:** Upon receipt of the signed contract, the client is informed of the date for stage 1 audit which is carried out at client's location to assess the client's food safety management system conform with all the requirements of the ISO 22000:2005 standard, to verify that the food safety management system is designed to achieve the organization's food

safety policy and objectives, to assess the capability of the food safety management system to manage compliance with statutory, regulatory and contractual requirements and to confirm the planned arrangements for the Stage 2 audit. The purpose of the initial audit is to assess the readiness of the management system for stage 2 audit, also called as 'Certification Audit' or 'Main Audit' and to plan for the Certification Audit. It is necessary that at the time of initial audit the requirements for certification as mentioned above are met.

4. **Stage 2 audit:** The Certification Audit is conducted within maximum 6 months of the 1st stage audit. The stage 2 audit takes place at client's site to confirm

    - The implementation, including the effectiveness of the client's food safety management system to the requirements of the ISO 22000:2005,
    - To verify that the information and evidence of conformity is achieved to all requirements of the food safety management system standard,
    - To assess the capability of the food safety management system to perform key activities, such as processes & activities including production methods, controls, PRP's, HACCP plans & procedures as well as competency of personnel managing and involved in the food safety functions, in conformity with ISO standard,
    - To assess the client's food safety management system in compliance with statutory, regulatory and contractual requirements;
    - To confirm that the client's food safety management system is effective in achieving the stated food safety policies and objectives

Based on the objective evidence of conformity or nonconformity to the requirements of ISO 22000:2005, the audit team leader in consultation with audit team members arrives at audit conclusion based on audit findings. If non-conformities are found during the audit, a corrective action request form is provided to client to comment on the nonconformities and a corrective actions plan will be decided in consultation with the client who should submit the corrective action report to, OneCert International upon closure of the non-conformities.. If there is any major non –conformity found during the stage 2 audit, the effectiveness of the corrective action will be verified by an on-site follow up audit.

5. **Issuing Certificate:** When all corrective actions agreed between the client and the audit team have been completed & Upon receipt of the full payment, the certificate is issued.

6. **Post certification activity**

   A. **Surveillance audits** Audit is conducted for first year (within Calendar Year from certification decision date) and second year (within Calendar Year from first surveillance audit this time frame for 2nd surveillance audit can be extended up to 60 days for some justified reasons (for example plant is shut down due to upgradation or major breakdown in plant/machinery, natural calamities etc.) to assess the clients continued compliance to the requirements of standard.

   B. **Recertification Audit** : Recertification audit is conducted in the same manner as of initial certification audit within every 3 years but before the expiry of certification.

   C. **Certification Maintenance:** The client is required to inform certifying agency promptly of any significant changes to its product(s) or services that impact the certified management system(s) or any other circumstances, which may affect the validity of its certification. For example: Change of site, Additional sites, Change of process, Change of ownership, Change of scope, change in number of employees etc.

International Certification Agencies

1. Control Union Certification
2. IFS – International Featured Standards
3. NSF International – Global Food Safety Initiative

Key Points

- The FSSAI or Food Safety and Standards Authority of India is the regulatory body for Food Safety for all FBOs in India under One Nation One Food Law.
- It lays down rules and guidelines mandatory for all food businesses.
- Obtaining licenses are mandatory for all FBO.
- Licenses are generated through online portal which shall shift from FLRS to FoSCoS w.e.f. Nov 2020. All licenses are generated within 60 days of complete application submission
- There are three level of licenses i.e. Registration, State license and Central license.
- FSMS ensure the safety of food through supply chain.
- PRPs provide basic environment and operating conditions for production of safe foodbut not specific to any FBO.

- HACCP focuses on *prevention of hazards* rather than finished product inspection
- Many certifications are desirable but not mandatory. However, it is necessary to develop trust in the system adopted by any company.

References

Anonymous, 2020. The main certificates of the food industry. (Source: btsa.com, retrieved 27 Nov 2020).

JoyceAnnemarie, 2019. Understanding the difference between PRP, OPRP & CCP – An Introduction.https://safefood360.com/2019/08/understanding-the-difference-between-prp-oprp-ccp-an-introduction/. Accessed 28.10.2020

Sharma, 2010. Postharvest Management and Processing of Fruits and Vegetables – Instant Notes. New India Pub Agency, New Delhi.

Index

C

G

H

I

J

K

L

M

N

O

P

T